"Photographed All the Best Scenery"

Jack Hillers's Diary
of the Powell Expeditions,
1871–1875

Volume nine of the University of Utah Publications in the American West,
under the editorial direction of the Center for Studies of the American West,
Don D. Walker, General Editor.

Jack Hillers working with his photographic equipment, Aquarius Plateau, Utah, 1872. James Fennemore photograph.

"Photographed All the Best Scenery"

Jack Hillers's Diary of the Powell Expeditions, 1871–1875

Edited by Don D. Fowler

University of Utah Press
Salt Lake City, Utah

Contents

Preface

One of the epic adventures of the nineteenth century American West was the exploration of the Colorado River in 1869 and 1871–72 by John Wesley Powell and his associates. The drama of piloting small boats at great risk through the rapids and canyons of the Green and Colorado rivers has held the imagination of Americans for a hundred years. The literature relating to the Powell expeditions is extensive. Powell's own popular account, though historically inaccurate, has been reprinted several times. Over the years historians have gathered and published the journals and letters of most of the expedition participants. One important diary, however, has remained unpublished until now. It is that kept by John K. (Jack) Hillers, who was a boatman on the second expedition and who later became the chief photographer for the United States Geological Survey and one of the great photographers of the nineteenth century.

Hillers's diary is an important addition to the historical literature of the Powell expeditions. It adds new information about the second expedition and its operations. It also includes sections describing Hillers's later work with the Powell Survey in 1874 and 1875. By publishing the diary with several examples of Hillers's photography, we hope to provide evidence for the claim that Hillers's work ranks with other great nineteenth century photographers. He was the first to photograph the Grand Canyon. He developed a process for making large photographic transparencies on glass which were used at a number of national and international expositions and fairs at the turn of the century. But most importantly, Hillers had a good eye for composition and detail. Many of his photographs are true masterpieces.

At the time the Powell expedition journals were being published in 1947–49, Hillers's diary remained in his family's possession. Later, Mrs. John K. Hillers, Jr., donated the diary to the Smithsonian Institution. In 1968 Mrs. Hillers kindly granted permission to me to publish the diary. Special thanks are due Mrs. Hillers for her kind permission and for permitting my wife and me to examine the Hillers family scrapbooks and memorabilia.

Since 1967 Catherine S. Fowler and I have been editing for publication a series of manuscripts, including the Hillers diary, relating to John Wesley Powell, the Powell expeditions, and the Bureau of American Ethnology. The manuscripts are on deposit in the Smithsonian National Anthropological Archives, Washington, D.C. The work was made possible by a grant from the National Endowment for the Humanities and a National Academy of Sciences Post-doctoral Research Fellowship at the Smithsonian in 1967–68. This support is acknowledged with gratitude. John C. Ewers, Margaret Blaker, Clifford Evans, and Joanna Scherer of the Smithsonian have been most helpful with the Hillers diary. Nell Carico of the United States Geological Survey, Marilyn Seifert of the Latter-day Saints Church Historian's Office, Everett L. Cooley of the University of Utah, Jack Haley of the University of Oklahoma, Rella Looney of the Oklahoma Historical Society, Vilate Hardy Gubler of La Verkin, Utah, and Juanita Brooks of St. George, Utah, have all supplied information to aid in annotating and verifying details of the diary. My thanks to each of them. The photographs made by Hillers and his contemporaries are reproduced here with the permission of the United States Geological Survey

Records Department of the National Archives (frontispiece and figures 1–13, 23–26, and 39) and with the permission of the Bureau of American Ethnology Collection of the Smithsonian National Anthropological Archives (figures 14–22, 27–38 and 40–44). Finally, especial thanks to my wife, Catherine S. Fowler, for editorial, critical, and general support.

I wish to express my appreciation to Princeton University Press for permission to quote from *Powell of the Colorado* by William Culp Darrah (copyright 1951 by Princeton University Press and 1969 by Princeton Paperback), p. 212; and to Dover Publications, Inc., for permission to quote from *Photography and the American Scene* by Robert Taft, pp. 289–91. Photographs on pages 16 through 19 are courtesy of William Culp Darrah, the Utah State Historical Society, the New York Public Library, the Bureau of American Ethnology Collection of the Smithsonian National Anthropological Archives, Mrs. Virginia B. Chase, Mrs. Mabel Powell Bradley, the New York Public Library, and Herbert E. Gregory respectively.

Introduction

In May 1871, John K. (Jack) Hillers chanced to meet Major John Wesley Powell in Salt Lake City. Powell and his brother-in-law Almon Harris Thompson were in the city to get their wives settled and to make final preparations for a second boat trip down the Green and Colorado rivers. One of the men scheduled to accompany the expedition could not go and Powell was looking for a replacement. Somehow Powell met Hillers, who was working in the city as a teamster. Powell liked the looks of the tall, red-haired Hillers and offered him the boatman's job. Hillers accepted.

The meeting marked the beginning of a lifetime association between Powell, Thompson, and Hillers. In subsequent years Powell was to become the director of both the Bureau of American Ethnology and the United States Geological Survey; Thompson was to become the chief topographer of the Geological Survey; Hillers was to become the chief photographer for Powell's Geological and Geographical Survey of the Rocky Mountain Region, and, in 1881, the chief photographer for the United States Geological Survey. Hillers was the first to photograph the Grand Canyon. Through the years he made many series of now classic photographs of the Indians and the geological formations of the Colorado Plateau and the American Southwest. His work ranks with that of other great nineteenth century photographers of the American West, including Timothy O'Sullivan, William H. Jackson, Charles R. Savage, and Christian Barthelmess.[1]

Hillers kept a diary of the events of the trip down the Green and Colorado rivers in 1871 and 1872, of some of his work for the Powell Survey in 1874, and of a trip to Indian Territory in 1875. The present volume presents Hillers's diary for the first time, together with a selection of his photographs made between 1872 and 1879. The volume is thus a contribution to the history of the Powell expeditions and the American West and to the history of American photography.

The Powell Survey

Beginning with the Lewis and Clark expedition in 1803–05, the United States government sponsored many expeditions to the American West to make maps and to collect data on resources and Indian tribes. Prior to the Civil War these expeditions were conducted by the United States Army Corps of Topographical Engineers.[2] After the Civil War, federal support for western exploration continued, but a new type of survey developed. With the exception of the Geographical Survey West of the One Hundredth Meridian, led by Captain George M. Wheeler of the United States Army, the new surveys were under civilian control.

[1] Robert Taft, *Photography and the American Scene: A Social History, 1839–1889* (New York, 1964); Maurice Frink with Casey Barthelmess, *Photographer on an Army Mule* (Norman, Okla., 1965); James D. Horan, *Timothy O'Sullivan: America's Forgotten Photographer* (New York, 1966); William H. Goetzmann, "Images of Progress, the Camera Becomes Part of Western Exploration," in William H. Goetzmann, *Exploration and Empire: The Explorer and the Scientist in the Winning of the American West* (New York, 1966), pp. 603–648.

[2] Goetzmann, *Exploration and Empire*, pp. 231–331; William H. Goetzmann, *Army Exploration in the American West* (New Haven, 1959).

2

Ultimately there were four federally sponsored surveys, including the Wheeler Survey, operating in the West. The three civilian surveys were the Geological and Geographical Survey of the Territories, led by Ferdinand Vandeveer Hayden; the Geological Exploration of the Fortieth Parallel, led by Clarence King, and the Powell Survey, which had several names but was best known as the Geological and Geographical Survey of the Rocky Mountain Region.[3] The Powell Survey was the smallest and the last on the scene, but it, and its director, were to have enormous impact on federally sponsored scientific research, land use practices, and conservation practices. In 1879 all four surveys were merged into the United States Geological Survey.

The story of Powell's survey and subsequent career as director of two major government scientific agencies, the Bureau of American Ethnology and the United States Geological Survey, is well known.[4] Here we will only sketch Powell's career in its relation to Hillers and the work Hillers performed under Powell's direction.

Powell emerged from the Civil War with one arm and the rank of major to become a professor at Illinois Wesleyan University. In 1867 he led a party of students and friends on an expedition to the Rocky Mountains.[5] In 1868 he led a second party to the same area, but in the fall he, with his wife and a few others, remained behind to explore the upper reaches of the Green River in western Colo-

[3] Goetzmann, *Exploration and Empire*; Richard A. Bartlett, *Great Surveys of the American West* (Norman, Okla., 1962).

[4] William C. Darrah, *Powell of the Colorado* (Princeton, 1951); Wallace Stegner, *Beyond the Hundredth Meridian: John Wesley Powell and the Second Opening of the West* (Boston, 1954).

[5] Elmo Scott Watson, ed., *The Professor Goes West: Illinois Wesleyan University Reports of Major John Wesley Powell's Explorations, 1867–1874* (Bloomington, Ill., 1954).

rado, eastern Utah, and southern Wyoming. During the winter Powell formulated plans to explore the Green and Colorado rivers in boats.[6] In 1868–69 only the Green River above the Uintah Basin and the lower Colorado below Black Canyon (the present site of Hoover Dam) were known and mapped. The area between, comprising most of the deeply dissected Colorado Plateau, was virtually unknown. Some brief attempts had been made to run the upper Green River as early as 1825, and rumors abounded of giant whirlpools and rivers disappearing underground — but there were few facts.[7]

Powell resolved to explore the river system. He gained support from several institutions in Illinois and on May 24, 1869, he and nine men in four boats set out from Green River Station, Wyoming Territory. On August 30, 1869, Powell and six men in two boats emerged from the foot of the Grand Canyon at the small Mormon settlement of Callville, Nevada. Along the way two boats were lost or abandoned and three men had left the party in the Grand Canyon. These men were later killed by Shivwits Indians.

The river trip made Powell a national hero — and gained him a congressional appropriation to continue his explorations. Much of the data collected during the first trip was lost and a shortage of supplies had forced the party to hurry. But Powell had learned that the rivers were navigable by properly constructed small boats. As difficult and fearful as the rapids and falls were in many places, Powell and his men had learned how to run them or portage around them.

Powell now planned a second trip, this time with supplies cached along the

[6] Ibid., p. 24.

[7] Don D. Fowler, Foreword, John Wesley Powell, *Down the Colorado: Diary of the First Trip Through the Grand Canyon* (New York, 1969), p. 14.

way by overland pack teams and with adequate scientific apparatus and time to use it. The 1869 trip had been an adventure into the unknown; the 1871 trip was a scientific expedition designed to collect new geological, hypsometric, and other data — and to photograph for the first time the fantastic Canyon Country.

In May 1871, Powell and Almon Harris Thompson, Powell's brother-in-law and chief topographer, arrived in Salt Lake City, met Hillers, and hired him.

Powell, Thompson, and Hillers arrived at Green River Station on May 16, 1871, to find the other members of the river party waiting for them. Hillers's diary begins on that date.

Following the second river trip Powell was able to gain continued congressional appropriations for his survey. He, Thompson, and others spent most of the 1870s studying and mapping the geology of Utah and Arizona and producing reports which remain classics in the field. Hillers accompanied the various field parties as photographer. Many of his photographs were used as the bases for the engraved illustrations by Thomas Moran and others in the geology reports produced by members of the Survey.[8]

Powell's interests included anthropology as well as geology. He had begun studying the Ute and Southern Paiute Indians of the Colorado Plateau in 1868 and became competent in their languages. Whenever he got the chance he recorded ethnographic and linguistic data.[9]

[8] E.g., the illustrations in Clarence E. Dutton, *Report on the Geology of the High Plateaus of Utah* (Washington, 1880), and in John Wesley Powell, *Exploration of the Colorado River of the West and Its Tributaries* (Washington, 1875).

[9] Don D. Fowler and Catherine S. Fowler, eds., "Anthropology of the Numa: John Wesley Powell's Manuscripts on the Numic Peoples of Western North America, 1868–1880," *Smithsonian Contributions to Anthropology* vol. 14 (1971).

In May 1873, Powell was appointed a Special Commissioner of Indian Affairs to look into a number of Indian problems in Utah, Idaho, northern Arizona, and Nevada. Together with George W. Ingalls, Powell met various Indian delegations in Salt Lake City and then traveled south meeting various Indian bands and delegations along the way.[10] Hillers met Powell in Kanab, Utah, and accompanied him to St. George, Utah, and Moapa and Las Vegas, Nevada. At these points, Hillers made a series of photographs of various Southern Paiute Indians. These photographs provide valuable ethnographic data on the Indians.[11] But some few of them are not true ethnographic records. Powell had collected some buckskin clothes from the Northern Utes of the Uintah Basin in 1868. These he brought with him to southern Utah and dressed some of the Southern Paiutes in them. In the photographs, one, and possibly two, Southern Paiute women are dressed in Northern Ute beaded buckskin dresses. In one photograph, the word "Colorado" and a museum accession number are clearly visible on the bodice of the dress (Figure 15). In some of the photographs the men are wearing feather headdresses. Such ornamentation was not indigenous to the Southern Paiute, who usually wore close-fitting caps. There is some evidence that the headdresses were made for the occasion under the direction of Ellen Thompson, Powell's sister. Some of the poses in the photographs are highly stylized nineteenth century "art" poses. There is also an element of cheesecake in some of the photographs: some Indian women were posed with one breast just visible.

[10] John Wesley Powell and G. W. Ingalls, *Report of Special Commissioners J. W. Powell and G. W. Ingalls on the Condition of the Ute Indians of Utah . . .* (Washington, 1874).

[11] See Julian H. Steward, . . . *Notes on Hillers' Photographs of the Paiute and Ute Indians Taken on the Powell Expedition of 1873*, Smithsonian Miscellaneous Collection, vol. 98, no. 18 (Washington, 1939), and Robert C. Euler, *Southern Paiute Ethnohistory*, University of Utah Anthropological Papers no. 78 (Glen Canyon Series no. 28) (Salt Lake City, 1966), Appendix I.

Although Powell recognized the ethnographic value of the photographs, he had other uses for them in mind as well, seeing them primarily as a source of income. There was a substantial market in the 1870s for stereographs for the stereoscopes found in most nineteenth century homes.[12] Powell and Beaman had made an arrangement to split the proceeds from the stereographs produced during the river trip. But when Powell and Beaman disagreed and Beaman left in January 1872, Powell bought out Beaman's interest. After Hillers became Survey photographer, he, Powell, and Thompson entered into an agreement to share the proceeds of the sale of stereographs. Powell received forty percent, Thompson thirty percent, and Hillers thirty percent. Powell took Hillers's share from the proceeds of Beaman and Fennemore negatives. There are few figures on the proceeds from these sales, but it is known that they totalled $4,100 for the first six months of 1874. Darrah notes that a standing joke in the United States Geological Survey in the late 1880s was that Powell had paid off the mortgage on his house in Washington through sales of the views.[13]

[12] Taft, *Photography and the American Scene*, pp. 17–21.

[13] Darrah, *Powell of the Colorado*, p. 182, n. 7. An itemized statement in the archives of the Office of the Secretary in the Smithsonian Institution lists the sum of $2,413.89 for photographic materials and the employment of Beaman in 1872. The statement includes $1,173.70 for equipment and supplies, $800.00 to Beaman "for services rendered," and $440.19 to W. H. Jackson. A separate itemized bill from Jackson, dated July 5, 1872, lists charges for printing 4,288 stereographs and for mounting 549 others between March 10 and July 5, 1872. ("Statement of Expenditures for Photographic Apparatus and Material, *and* also for the Employment of a Photographic Artist, for the Exploring Expedition of the Colorado River of the West," J. W. Powell to Joseph Henry, December 12, 1872, Henry Papers, Archives, Office of the Secretary, Smithsonian Institution.)

In addition to selling them, Powell used stereographs in other ways. During appropriation hearings and at other critical times in Congress, Powell sent sets of views to various key congressmen. In 1877 Powell was particularly anxious about the continuation of his appropriation. The

After 1873 Powell established a regular office in Washington for his Survey. Hillers and other Survey employees thereupon spent part of the year in Washington. Sometimes Hillers accompanied Powell on lecture tours.[14]

In 1875 Powell was asked to direct the collection of Indian artifacts, information, and photographs for the Smithsonian Institution exhibit at the 1876 Philadelphia Centennial Exposition. As a part of this effort, Powell sent Hillers to Oklahoma, then Indian Territory, to take pictures of various Indian tribesmen. The last portion of Hillers's diary is devoted to this trip.

In 1879 Powell succeeded in getting the Bureau of Ethnology (after 1894, the Bureau of American Ethnology) established as a part of the Smithsonian Institution, with himself as director.[15] Hillers was hired as Bureau photographer. One of Powell's first acts was to send Hillers, together with James Stevenson and Frank Hamilton Cushing, to Arizona and New Mexico to survey archeological

Rocky Mountain Survey letter books show that between January and March of 1877, sets of views were sent to thirteen Congressmen and other politically powerful men including Simon B. Cameron, a senator from Pennsylvania and previously Lincoln's secretary of war, and John Sherman, senator from Ohio and secretary of the treasury under Hayes.

During the period 1871–75 several sets of views were made, some scenic, some of Indians. Sets include "Views of the Green River," "Views on the Colorado River," "Views on the Sevier River," each with sub-series, and sets of views of Indians of the Colorado River, the Navajo and the Hopi and Zuni. A photographic catalog (in the Smithsonian National Anthropological Archives), the "Catalog of Negatives, River, Land and Ethnographic, 1871–1876," lists 131 negatives by Beaman, 71 by Fennemore, 27 by Fennemore and Hillers, and 368 by Hillers. Sets of the views in the Denver Public Library indicate that some were published by J. F. Jarvis (who for a time worked for W. H. Jackson), in Washington, D.C., and others by the William B. Holmes Company in New York City. Some views were copyrighted by Powell in 1874. Others, issued after 1879, have the statement, "Smithsonian Institution, Bureau of Ethnology, compliments of J. W. Powell," printed on the back.

[14] Darrah, *Powell of the Colorado*, p. 213.

[15] Ibid., pp. 254 ff.

ruins and photograph the Pueblo Indians.[16] Several of Hillers's best pictures derive from this trip.

Powell had had a behind-the-scenes hand in the formation of the United States Geological Survey, also formed in 1879. In 1881, the director, Clarence King, resigned and Powell was appointed to replace him, an appointment he took on in addition to his directorship of the Bureau of Ethnology. Hillers was transferred to the Survey payroll (which always had more money than the Bureau) and became the chief photographer of the Survey at a salary of $1,800 per year.[17] He remained in that capacity until his retirement in 1900. Between 1881 and 1894, when political pressures forced him to resign as Director of the Survey, Powell ran both the Survey and the Bureau from the same office. Hence, Hillers was usually available to take portraits of visiting Indian delegations for the Bureau.

Hillers's Life

There is little biographical data about Hillers's early life.[18] He was born in Hanover, Germany, in 1843 and came to the United States at the age of nine. At the beginning of the Civil War he enlisted in the New York Naval Brigade, but later

[16] John Wesley Powell, "First Annual Report of the Director of the Bureau of Ethnology (for the fiscal year 1879–80)," *Bureau of Ethnology, First Annual Report* (Washington, 1881), p. xxx; Raymond S. Brandes, *Frank Hamilton Cushing: Pioneer Americanist*, Ph.D. diss., University of Arizona, Tucson (University Microfilms 65–9951), pp. 25–35.

[17] United States Civil Service Commission, *Official Register of the United States, Directory, 1816–1959* (Washington, 1907–).

[18] This brief biographical sketch is adapted from an obituary in the Washington, D.C. *Star*, November 16, 1925; from William C. Darrah, "Beaman, Fennemore, Hillers, Dellenbaugh, Johnson, and Hattan," *Utah Historical Quarterly* 16–17 (1948–49): 495–97; and from personal interviews with Mrs. J. K. Hillers, Jr., of Washington, D.C.

transferred to the army. He saw action at Petersburg, Cold Harbor, and Richmond. At the end of the war he re-enlisted and served in various Western garrisons.

In 1870 Hillers resigned from the army to accompany his brother Richard to San Francisco. Returning eastward in 1871, he stopped in Salt Lake City and went to work as a teamster. There, as we have seen, he chanced to meet Powell and Thompson and signed on their river expedition. Hillers was well liked by all. His capabilities, sense of humor, and evenness of temper earned him the nickname of "Jolly Jack," though he was most often called "Bismark" (*sic*) for his German background. Hillers was especially close to Powell. As Darrah has written, "Hillers' buoyant spirit and, at times, ribald sense of humor, endeared him to the Major. He was the one man who could be breezy and flippant with his chief. Hillers alone dared write to the Major, 'Love to Mrs. Powell and kisses for the girls. . . .' " [19]

Hillers's transition from boatman and handyman to chief photographer of the Powell expedition took place during the 1871 river trip and in the following year. Powell had hired E. O. Beaman, a professional photographer from New York City, to make photographs during the river trip. Beaman's equipment—large format camera, portable darkroom, chemicals, and large glass negatives for the collodion wet-plate process then used—weighed nearly a ton. Powell appointed his cousin, Walter Clement (Clem) Powell, Beaman's assistant.

During the river trip Hillers became interested in photography. Frederick S. Dellenbaugh, a member of the expedition, in a letter to Robert Taft, the historian of early American photography, explained how Hillers became a photographer:

> Jack Hillers was engaged by Powell for the second trip, in Salt Lake
> City, to pull an oar and to help generally. He knew nothing then about

[19] Darrah, *Powell of the Colorado*, p. 212.

photography, but he became much interested as we went on down Green River.

When we had arrived at the mouth of the Green . . . Hillers asked me about photography — about the chemical side. I explained about the action of light on a glass plate coated with collodion and sensitized with nitrate of silver, the bath to eliminate the silver in hypo, and so on.

"Why couldn't I do it?" he said. I replied that he certainly could, for he was a careful, cleanly man, and those were the chief qualities needed. I advised him to offer his help to Beaman whenever possible . . . and perhaps Beaman would let him try a negative. He did, and in two or three weeks had made such progress that he overshadowed Clement Powell.

When Beaman left at the end of our first season of river work, the photography fell on Clement. He made a trip with Hillers as assistant and returned with nothing. He declared Beaman had put hypo in the developer. Beaman . . . said he had not done so — that the trouble was Clem's carelessness.

Meanwhile, Powell had sent to Salt Lake City to see if a photographer could be found there who would come down [to Kanab, Utah,] and who would go through the Grand Canyon with us the next summer. James Fennemore came. He was an excellent photographer and a genial fellow. . . . He was good to Hillers and gave him much instruction, with the result that Hillers became expert in the work, became assistant, in fact. . . .

Hillers did some excellent work with Fennemore's guidance. When we were preparing to enter the Grand Canyon in the summer of 1872 Fennemore was taken sick . . . [and] could not proceed. And we could not find anyone else who desired to do the Grand Canyon with us. This left us with only seven men. One boat had to be left behind. . . . Well what I am coming to is this; there was nothing for it but to make Jack

Hillers photographer-in-chief. He was equal to the job. In spite of enormous difficulties, great fatigue, shortage of grub, etc., he made a number of first class negatives.[20]

Hillers made several field trips in the 1880s. While in the field he was, as always, resourceful. Captain John Gregory Bourke, the famed ethnologist and aide-de-camp to General George Crook, ran into Hillers and a Survey crew in northern Arizona near the Mormon settlement of Sunset in the summer of 1881. Libations were clearly in order. As Bourke reports, "Hillers reappeared with a mixture of ginger and whiskey. The ginger was all right, but the whiskey, from the Mormon town of Brigham, was as vile as Arizona could produce." [21]

After 1885 increasing administrative duties kept Hillers in Washington most of the time. In 1886 he received a raise in pay — to two thousand dollars per year, a sum he continued to earn until his retirement. On May 1, 1900, Hillers retired from the Survey, but continued to work on a per diem basis until 1919.[22] In September 1902, Hillers had the sad duty of serving as a pallbearer at Powell's funeral.[23] The close relationship between the two men had continued until the end.

John K. Hillers died in Washington, D.C., in 1925 and was buried in Arlington National Cemetery, near the graves of Powell and Thompson.

[20] F. S. Dellenbaugh to Robert Taft, November 19, 1932, printed in Taft, *Photography and the American Scene*, pp. 289–91.

[21] John Gregory Bourke, *The Snake-dance of the Moquis of Arizona, Being a Narrative of a Journey from Santa Fé, New Mexico, to the Villages of the Moqui Indians of Arizona . . .* (New York, 1884), p. 354.

[22] Civil Service Commission, *Official Register*, p. 78.

[23] Darrah, *Powell of the Colorado*, p. 397.

Hillers's Photographs

Jack Hillers made several thousand negatives of anthropological and geological subjects in his twenty-nine years with the Powell Survey, the Bureau of Ethnology, and the Geological Survey. Many of these are deposited in the National Archives and the Smithsonian National Anthropological Archives in Washington, D.C. The Denver Public Library also has a large collection of Hillers's photographs and stereographs.

Hillers was the first man to photograph the Grand Canyon and most of the high plateaus of central and southern Utah.[24] His ethnographic series of photographs of Southern Paiute Indians, although some of his models are posed in artificial attitudes, are a valuable source of information, as are his photographs of Zuni and various Oklahoma Indians.

Hillers pioneered the making of large photographic transparencies on sheets of glass up to 4×5 feet in size. Some of these were hand colored. They were used in several of the Smithsonian Institution and Bureau of Ethnology exhibits at national and international fairs and expositions.

Hillers was first and foremost a craftsman. His photographs are technically excellent, a remarkable achievement given the wet-plate negative process in use until the 1880s. But in many of his photographs he went beyond technical excellence. In composition and content many of his photographs are works of art.

Hillers's Diary

Most of the known diaries and journals of the Powell river expeditions of 1869

[24] Stegner, *Beyond the Hundredth Meridian*, p. 268.

and 1871–72 have previously been published.[25] Hillers's diary was deposited in the Bureau of American Ethnology archives (now incorporated into the Smithsonian National Anthropological Archives) in 1952 by Mrs. J. K. Hillers, Jr., Hillers's daughter-in-law. It consists of two small leather-bound books and a number of loose sheets. In 1952 Matthew W. Stirling, then Chief of the Bureau of American Ethnology, collated and organized the loose sheets and had a typescript copy made. In 1968 the present editor checked the typescript against the original diaries. The script is small and sometimes difficult to read, some words being illegible.

Hillers apparently started his diary some weeks after the beginning of the river trip. Stephen Vandiver Jones's diary for August 9, 1871, contains the following

[25] The journals of George Y. Bradley, John C. Sumner, and John Wesley Powell, some letters of O. G. Howland, and other related materials from the 1869 river expedition (edited by W. C. Darrah) are found in *Utah Historical Quarterly* 15 (1947): 1–148. See also O. Dock Marston, "The Lost Journal of John Colton Sumner," *Utah Historical Quarterly* 37 (1969): 173–89.

The documentation of the second river trip of 1871–72 is extensive. It includes the journal and letters of Francis Marion Bishop, ed. Charles Kelly, *Utah Historical Quarterly* 15 (1947): 154–253; the journal of Stephen Vandiver Jones, ed. Herbert E. Gregory, *Utah Historical Quarterly* 16–17 (1948–49): 19–174; the journal and newspaper articles written by Walter Clement Powell, ed. Charles Kelly, *Utah Historical Quarterly* 16–17 (1948–49): 257–478; the diary of Almon Harris Thompson, ed. Herbert E. Gregory, *Utah Historical Quarterly* 7 (1939): 1–140; John Wesley Powell, "John Wesley Powell's Journal: Colorado River Exploration, 1871–1872," ed. Don D. Fowler and C. S. Fowler, *Smithsonian Journal of History*, vol. 3, no. 2 (1968), pp. 1–44; Frederick S. Dellenbaugh, *A Canyon Voyage: The Narrative of the Second Powell Expedition Down the Green-Colorado River from Wyoming, and the Explorations on Land, in the Years 1871 and 1872* (1908, reprint ed., New Haven and London, 1926); and E. O. Beaman, "The Cañon of the Colorado, and the Moquis Pueblos," *Appleton's Journal* 11 (1874): 481–84, 513–16, 545–48, 590–93, 623–26, 641–44, 686–89.

Powell's own published version of the two river trips, written as if all events took place in 1869, is contained in J. W. Powell, *Exploration of the Colorado River of the West*. Other related materials are found in Watson, *Professor Goes West*; Paul Meadows, "John Wesley Powell: Frontiersman of Science," *University of Nebraska Studies*, n.s., 10 (1952): 1–35; and Lindsey G. Morris, "John Wesley Powell: Scientist and Educator," *Illinois State University Journal*, vol. 31, no. 3 (1969), pp. 3–46.

passage: "Hillers has procured a book and spent the time today in writing up from the start. Believe each member of the party is keeping a journal with the exception of Hattan." [26]

On August 9, 1871, Hillers had just returned from the Uintah Indian Agency with supplies for the river party. He may have acquired the book at the agency. After the river party reached Kanab, Hillers apparently mailed sections of the diary to his brother Richard in San Francisco, and later sent other sections to him. An undated page in the materials contains the note, "Dick, don't lose any leaves out of this, be careful. You can amuse yourself reading it then put it in my trunk."

The diary covers the period from May 16, 1871, through October 26, 1872 (with some breaks), as well as five days (September 11–15) probably in 1873, and the period May 1 through June 10, 1875. Included with the 1875 material is a portion of an undated letter to Hillers's brother describing a meeting with a charming Indian girl in Indian Territory. Unfortunately, the last section of the letter is missing and we do not learn of the outcome of the meeting.

In editing the diary for publication care has been taken to retain Hillers's original spelling. Editorial inserts in brackets correct or clarify proper names as necessary.

[26] Stephen Vandiver Jones, "Journal of Stephen Vandiver Jones," ed. Herbert E. Gregory, *Utah Historical Quarterly* 16–17 (1948–49): 57–58.

The
Members
of the

John Wesley Powell (1834–1902). Powell was born in New York. In the Civil War he achieved the rank of major, a title he retained throughout his life. He lost his right arm at the battle of Shiloh. After the war he became a professor, leading natural history expeditions to the Rocky Mountains in 1867–68, and in 1869 leading a party on the first daring descent of the Green and Colorado rivers. In 1879 he became the first Director of the Bureau of (American) Ethnology of the Smithsonian Institution. From 1881 to 1894 he was also Director of the United States Geological Survey. Powell died in 1902, internationally known as a scientist, administrator, and conservationist.

Almon Harris Thompson (1839–1906). Thompson was called Prof. Born in New Hampshire, Thompson moved with his family to Illinois where he graduated from Wheaton College in 1861. He married Ellen Powell, J. W. Powell's sister. Thompson joined J. W. Powell on the latter's reconnaissance of southern Utah and Arizona in 1870 in preparation for the second river expedition. Thompson became the chief topographer of the expedition and the *de facto* field supervisor. He was the chief geographer of the United States Geological Survey from 1879 until his death in 1906. The map of the Colorado River area was largely Thompson's work.

Frederick Samuel Dellenbaugh (1853–1935). Dellenbaugh was nicknamed Fred, Rusty, and Sandy by members of the expedition. He was a distant relative of Almon Harris Thompson and was only seventeen when he hired on as an artist for the Powell expedition. In 1873 he left the expedition and traveled in the West and Europe. Later he joined the Harriman expedition to Alaska and Siberia and made trips to South America and the Caribbean. He was a founder of the Explorers Club, wrote two books on the Powell expedition, and was responsible for collecting and preserving many of the diaries of members of the Powell expedition.

John K. (Jack) Hillers (1843–1925). Jack Hillers was born in Hanover, Germany, and came to the United States in 1852. He entered the Union Army during the Civil War and remained until 1870. In 1871 he chanced to meet John Wesley Powell in Salt Lake City and signed on as a boatman for Powell's second expedition down the Colorado River. Hillers later became Powell's chief photographer, serving in that capacity until 1881 when he became chief photographer of the United States Geological Survey. He held that post until 1900 when he retired. He continued on a part-time basis until 1919. He died in 1925 and was buried in Arlington National Cemetery.

Francis Marion Bishop (1843–1933). Bishop was nicknamed Cap and Bish. He was born in New York. Running away from home at age seventeen to join the Union Army, Bishop was promoted to first lieutenant for bravery at the first battle of Bull Run and at Antietam. He was badly wounded at Fredericksburg in 1862 but was eventually promoted to captain. Bishop entered Illinois Wesleyan University where Powell hired him as topographer for the second expedition. Bishop in 1872 settled in Salt Lake City. He taught at the University of Deseret and in 1877 opened an assay office. In 1873 he married Alzina Pratt, a daughter of the Mormon leader Orson Pratt.

Walter Clement Powell (1850–1883). Nicknamed Clem by the expedition members, Powell was J. W. Powell's first cousin. His parents died when he was six and he was raised by J. W. Powell's family. Clem was supposed to learn photography and become a permanent member of the expedition. But he never mastered the art and was replaced by Hillers. Clem returned to Illinois and entered the pharmacy business. He later married Mary Breasted, the sister of the famous orientalist and historian James Henry Breasted. He moved to Omaha, Nebraska, and became a prominent wholesale druggist. He died suddenly in 1883.

**Powell
Expedition,
1871–1872**

John F. Steward (1841–1915). Steward was nicknamed Ford, Sergeant, and Truthful James by members of the expedition. He was born in Illinois. Steward joined the Union Army in 1862 and in 1864 met J. W. Powell; the two men collected fossils in the trenches around Vicksburg during the long siege of that city. Powell later hired him for the second river expedition. Many of the geological sections in Powell's later works on Colorado River geology are credited to Steward. The exertions of the expedition exacerbated Steward's war wounds and he returned to Illinois in 1872. He later became an executive of the International Harvester Company.

E. O. Beaman. Little is known about Beaman's life. He apparently began working as a professional photographer after the Civil War. Powell hired him as a photographer on the recommendation of the E. and M. T. Anthony Photographic Supply Company of New York. He remained with the expedition until January 1872 when he and Powell had a disagreement and Beaman resigned. He subsequently made a trip to the Hopi mesas and published articles on the river trip and his Hopi trip in *Appleton's Journal.*

Andrew J. Hattan (1841–1919). Hattan was nicknamed Andy and General. Hattan first met Powell during the Civil War. He later joined the expedition as cook and boatman. In 1872 he returned to his home in Ohio where he worked as a farmer and teamster for the rest of his life.

Stephen Vandiver Jones (1840–1920). Jones was called Deacon by members of the expedition. Jones was born in Wisconsin and later moved with his family to Illinois. In 1870 he became principal of the Washburn, Illinois, school where he met Almon Harris Thompson, Powell's brother-in-law. Through Thompson, Jones was hired as an assistant topographer. He left the expedition in late 1872 and returned to his teaching duties in Illinois. In 1883 he moved to Dakota Territory where he became an outstanding attorney, helping to frame the constitution and legal code for South Dakota. He remained in that state until his death in 1920.

James Fennemore (1849–1941). Fennemore was born in London and presumably came to Utah as a Mormon convert. He was working in Charles R. Savage's photographic gallery in Salt Lake City when he was hired by Powell to replace Beaman. Early in 1872 he joined the expedition but ill health forced him to resign prior to the 1872 trip through Marble and Grand canyons. His most important contribution was teaching Hillers photography. Fennemore later opened his own photographic gallery. He is known for his portraits of Brigham Young and of John D. Lee seated on his own coffin prior to his execution for his alleged role in the Mountain Meadows massacre.

Frank Richardson. Richardson was called Little Breeches by the expedition members. Very little is known about this man. Powell sent him home once the river party reached Brown's Park, apparently because he was not strong enough for the trip.

Jack Hillers's Diary
of the Powell Expeditions,
1871–1875

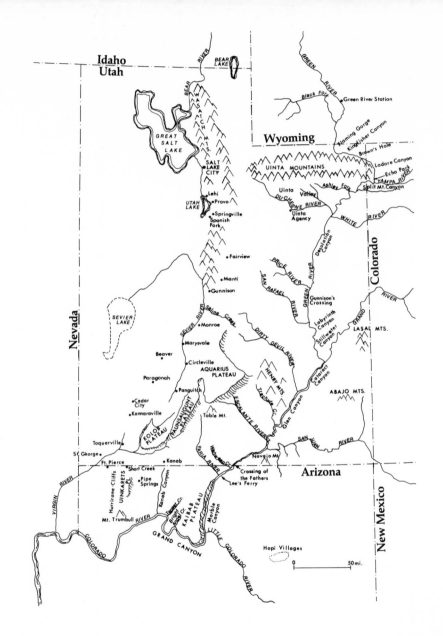

A Map of the Green and Colorado rivers,
1871–1872

22

1871

John K. Hillers

U.S. Survey of Colorado River 1871

Green River, May 16, [18]71

Left Salt Lake with Major Powell and Prof. Thompson at 5 a.m. to accompany them on the Exploring Expedition of the Green and Colorado Rivers. Got here at 6 a.m. where I found the rest of our party. The whole party stands as follows: Major [John Wesley] Powell, Geologist; Prof. [Almon Harris] Thompson, Astronomer and Topographer; [Francis Marion] Bishop (Cap) and [Stephen Vandiver] Jones (Deacon), Assistents; [John F.] Steward, Assistent Geologist; [E. O.] Beaman, Photographer; [Walter] Clement Powell, Assistent; [Frederick Samuel] Dellenbaugh, Artist; Frank Richardson, Barometrition; [Andrew J.] Hattan (General), Cook.

May 17. After breakfast the Major, Steward and myself climbed Fish butte, found some nice specimens of fossil Fish after a couple of hours search returned to camp. After dinner helped to finish boats.

May 18, 1871	May 18. At the boats again, covered them with canvas.
May 19, 1871	May 19. Painted canvas and made battings for cabins.
May 20, 1871	May 20. Had a row across the river, broke an oar on a gravel bank. Returned and had our pictures taken [Figure 1]. "Emma Dean", Maj. Powell, Jones, Helmsman, J. K. Hillers, Stroke, F. S. Dellenbaugh, Bow., "Nellie Powell," Prof. Thompson, Helmsman, Bishop, Bow., Steward, Stroke, Richardson, "Super Cargo," "Cañonita," Beaman, Helmsman, Hatten, Stroke, Clem Powell, Bow.
May 21, 1871	May 21. Loaded boats and got every [thing] ready for a start. Major bought an arm chair which I strapped on the centre cabin for him to sit in in order to have a better view of the River as we go.
May 22, 1871	May 22. After a heavy breakfast which Mr. [Jacob] Fields, the trader at this place, had provided for us, we repared to our boats, coiled our ropes, and made everything ship shape. We took our oars, at 10 a.m. we shoved off from shore with three hearty cheers for the people of Green River City — (Its population consisting of about 100 persons, whites, mongolian, and copper collored.), which was returned from the same on shore. Down we went at the rate of four miles an hour, but had not gone far when the Dean struck a bank in the centre. Her crew got out and dragged it over. The Nellie and Canonita followed suit. All got off safely. The Deacon broke his steering oar, the only damage done. Camped on the left bank

for dinner and observation. Pulled out again at two, nothing of interest this after noon. Camped on the right bank in an old hut belonging to some old trapper.[1]

May 23. Rain this morning, wound up with a snow storm which left us under cover of the old hut enjoying the warmth of a huge fire. After dinner pulled out very cold and blustry. Camped on the left bank.

May 24. Woke up this morning with frost on our blankets. Pulled out again. Camped for dinner under the walls of a high cliff. Immediately after pulling out ran aground, jumped out and soon floated down merrily. Camped opposite Black Fork, a stream of about 75 feet, very muddy.

May 25. Pulled out at 7 a.m., ran along smoothly. Scenery fine. Camped for dinner on left bank. After dinner passed an island on which we saw a dear. Major fired but missed, landed, and got three. One swam the river and thereby escaped the fate of his comrades. Camped on the left bank.

May 26. Pulled out in the morning. Saw some Beaver. Major shot twice but missed. Camped on the right bank after a short run for the rest of the day. Jerked our venison. Steward made geological sextions. Beaman took pictures. My friend Fred fixed his sketches.

[1] According to Stephen Vandiver Jones ("Journal of Stephen Vandiver Jones," ed. Herbert E. Gregory, *Utah Historical Quarterly* 16–17 [1948–49]: 24), the cabin was owned by "a white man who keeps the ferry at the [Green River] station."

May 27, 1871

May 27. This morning I was transferred to the Nellie Powell in place of Steward, who had not quite finished his work making sextions, while the Professor was anxious to make a certain point by 9 a.m. in order to get longitude and latitude. We had a hard pull of it but made it in time. We entered a mountain of Vermillion Red Sand Stone; it is called Flaming Gorge. Camped on the left bank in a beautiful cotton [wood] grove.

May 28, 1871

May 28. Water bright and clear. Major, Steward, and Richardson climbed the opposite bank to geologise the country. Richardson left them, for which he was nearly getting five days rations and an open road to Green River City. I fished, caught some very fine fish which we had for supper.

May 29, 1871

May 29. Pulled out again this morning. River snake-like.[2] After making several twists we came to one almost doubling and just at the centre we had our first rapid. Boats riding fine. Walls perpendicular, in some places, over hanging. Camped on the left bank for dinner, spent the rest of the day here. Prof., Steward, Bishop, Jones and Fred went across the river for Observation and Geological work. Major, Beaman, Clem and myself went up the river for pictures of the rapid and surrounding sceneries. Dropped down about 4 miles and camped near Kingfisher Creek on the right bank.

May 30, 1871

May 30. This morning the Major and myself climbed out, I for game, the Major for geological work.[3] Struck the Creek about 2 miles from camp. We tried

[2] The party was moving into Horseshoe Canyon.
[3] "Today with Bismark [Hillers] I climbed mountain S[outh] of Kingfisher Creek" (John Wesley Powell, "John Wesley Powell's Journal: Colorado River Exploration, 1871–1872," ed. Don D. Fowler and C. S. Fowler, *Smithsonian Journal of History*, vol. 3, no. 2 [1968], p. 9).

the trout first, but they being so stubborn would not take our decoy flys, so we started for the hills. Climbed up some 2000 feet. No sign of dear but got a splendid view of Kingfisher Creek for some 20 miles we could see a serpent form coming down through Cotton Wood Groves. Got back to camp tired.

May 31. Had a good rest last night, pulled out early, passed through King-fisher Cañon, ran a rapid, camped for dinner on the right bank. Beaman took some fine pictures of Kingfisher Creek and Cañon. Ran another rapid after dinner camped early on the right bank head of Red Cañon. Fred and myself at the oars, Hatten steering pulled Steward over on the opposite bank, the edge being thickly lined with willows, the boat got caught in a counter current flung her into the Willow brushing Steward off into the water, but we pulled him out in a saturated state.

June 1. Stayed in camp repairing and washing my unmentionables. While at work I can hear the dull roar of rappids, which means business tomorrow.

June 2. Got up early and immediately after breakfast we pulled out. After running a mile we came to the Rappids which had forwarned Business, and it was just "Biss" you bet. We got into a succession of "Raps" the boats half filling with water. At one place the river made a right angle turn, and just at the centre point a rappid. The Dean in running this almost filled, struck the rocky wall which tore out my oar lock. We would not have had this colision had it not been for the Major watching the Nell which was in a place which seemed disasterous. With only the use of two oars, we had hard work getting her ashore, as it was we had no time to spare for we were at the edge of another rappid. All hands jumped out to save the

May 31, 1871

June 1, 1871

June 2, 1871

27

boat from striking on sunken rocks, pulled her ashore, and bailed out. The "Nell" struck a rock close ashore rolled over, the men jumping on a flat rock. Richardson being excited would have been taken under had it not been for Prof. who pulled him off. The boat as soon as she was relieved of her deckload of human freight righted up again. She stove her sides some, but not enough to hurt. Damage was repaired in 30 minutes. "Canonita" came through all o.k. Passed two Creeks coming in on the right which we named Kettle and Compass, in commemoration of a kettle and compass we lost when the Nell turned over. Camped on the right bank for dinner. Beaman, Clem, Richardson and myself went up some ways to get a picture of the creeks, but could not get to it, the brush being too thick. Pulled out after dinner, ran several more rappids, and camped on right bank on the same spot where the Major had camped on the same day years ago.[4] Built a huge fire. Here we made our first portage, carrying the provisions for ¼ mile letting the boats down by line. After supper hung our clothing out to dry by the fire. Turned in early having done a big days work.

June 3, 1871

June 3. In camp this day. Fred caulking Bulkheads, and done his own washing for the first time in his life. The rest being all old soldiers know how to handle the "Kays of the Pianar" but I must give him credit, he done it well. Major, Beaman, Clem and myself took pictures, or rather Beaman took them, and we carryed the instruments.

June 4, 1871

June 4. This morning let the boats down by line 40 yards for this, when we got in to pull them through more rappids. After running several we came to one

---

[4] See George Young Bradley, "George Y. Bradley's Journal," ed. William C. Darrah, *Utah Historical Quarterly* 15 (1947): 33–34.

where we had to let down by line, after passing the rappid we got in to pull through more, in one of which Mr. Hook got drowned two years ago while attempting to follow the Major.[5] Camped for dinner on the left bank. After dinner concluded to stay all day. I muffled my oars with leather to keep them from slivering where the[y] work in the Locks.

June 5. Pulled out again this morning, ran more rappids but none very bad. Camped on the right bank for dinner. Prof., Beaman, Clem, Richardson and myself took some pictures of two Creeks coming in on the right, Francis & Cactus.[6] After dinner we pulled out with the Dean, looking out for Ashleys Fall, leaving the Nell and Canonita behind, on account of barometrical observation and for Beaman to finish his pictures. We had not many miles to pull when we could hear the dull roar of the Fall and soon after we came in sight. Landed just above, unpacked our boat and waited the coming of the other boats which soon made their appearance. They soon followed suit unloading. Carried our provision and baggage over the rocks first, then we got the Dean out of water and carried her over, no easy task. Men worked hard and with a will. Her keel and sides were much injured by dragging her over the rocks. She being so badly scraped, the Major concluded to let the other two down by line, so we took the Canonita next. We fastened a line to her stern and another to her bow. Down she went. When halfway over the fall

[5] "One rapid where Theodore Hook . . . was drowned in 1869 . . . gave us no trouble" (Frederick S. Dellenbaugh, *A Canyon Voyage: The Narrative of the Second Powell Expedition Down the Green-Colorado River from Wyoming, and the Explorations on Land, in the Years 1871 and 1872* [1908, reprint ed., New Haven and London, 1926], p. 25). Dellenbaugh notes that Hook was buried nearby; cf. Almon Harris Thompson, "Diary of Almon Harris Thompson," ed. Herbert E. Gregory, *Utah Historical Quarterly* 7 (1939): 15.

[6] Now called Trail Creek and Allen Creek respectively (Jones, "Journal," p. 31, n. 13).

we had to let go of her stern line in order to let her swing. While swinging around she got some heavy thumps on sunken rocks which filled her standing rooms with water, no other damage being done. The Major seeing how barely she escaped being knocked to pieces, concluded to try the first experiment with the Nell, which was done in about an hour. Camped for the night on the rocks. The Major named this Fall [Figure 2] after an old hunter who had placed his name on a rock. fac semela ASHLey The Major tells us that we will have smooth water until we reach LaDore.

June 6, 1871

June 6. Started out this morning anticipating smooth water but had not gone far when we heard a noise resembling a rappid, but of course having been told that we would have smooth water, we thought nothing of it. But all of a sudden turning an angle we found that little rappid round the corner, but got through it all right. Down the river we went meeting rappids after rappids, some of them as swift as 15 miles an hour. At dinner some one of the boys asked the Major if we would have any more smooth water, when he answered "well about the same". We partook of our meal on the left bank. Before reaching dinner station, Major, Beaman, Clem Richardson and myself went up a beautiful creek, took pictures of some falls. Major and Richardson left us to pull down some distance. Took Hatten out of the Canonita to prepare dinner. When they got there, leaving me in his place, Mr. Beaman placed Clem on one side of the fall and myself on the other. Satisfied with our success we followed, we called this Snow Creek. After dinner ran some 10 miles of about "the same" as we had in the fornoon when all of a sudden we opened out into a beautiful park. Major called it Red Canon Park. We camped on the right bank under two large Pine Trees. Took out our Ration to dry them. Caught a mess of fish.

June 7. In camp, drying ration. Washed and mended my clothing, half soled the Major's breeches. Only four of us in camp, rest went to climb a mountain some 4000 feet high. They returned at night, called it Mount Lena, after a bright flower of Cap Bishop's fancy.[7]

June 8. Pulled out this morning. A little more of "the same" when we came to an open space. Major called it Browns Park. On the left hand we saw some herders [Figure 3]. Landed and found that they had some mail for us. Camped for the rest of the day under a very peculiar shape Cotten tree. It resembled the horns of a deer. This park used to be called Browns Hole. The fact is, when a man "squats" on a piece of ground out here, his mud cabin is generally called his Hole. A man by the name of Brown, an old hunter, used to have his Hole here and hence its name. Several thousand cattle are kept here to fatten and then to be driven to California for consumption. This part is a splendid grazing country that is along the river bank.

June 9. I was surprised this morning to find that Richardson was to leave the Expedition.[8] The Major says that he is not strong enough to stand the trip, so he will leave us the day after tomorrow with Mr. Harroll [Harrell] for Green River City. Some cattle were drove up from which one was selected for our use. We gave them flour and sow belly in return. A greaser lassood one, a nice young heiver

[7] Lena was Bishop's sister (Francis Marion Bishop, "Capt. Francis Marion Bishop's Journal," ed. Charles Kelly, *Utah Historical Quarterly* 15 [1947]: 170).

[8] Interestingly, Powell ("Colorado River Exploration, 1871–1872," p. 13) makes no note of Richardson leaving the party, though other members do (Bishop, "Journal," p. 170; Jones, "Journal," p. 35).

about 3 year old, made a rather bungling job. I should advise him to go to San Francisco and there hire with the dog catchers. They would soon teach him how to throw a Lasso.

June 10, 1871

June 10. Wrote two letter, one for each brother in Frisco. Mr. Harroll will mail them at Green River. He ate supper with us, a Texan, and quite a gentleman.

June 11, 1871

June 11. Everybody had his little tricks ready. Mr. Harroll provided a fine riding animal for Frank R. and a pack mule to carry some nagitives for Mr. Beaman to be sent to Mrs. Powell at Salt Lake. Everything being ready the boys all shuk hands with Frank. He felt bad about leaving, the tears were in his eye but failed to shed any. I felt bad about him leaving. Immediately after we shot out like an arrow. Ran a few rappids. Low grounds for a few miles on each side, only here and there where the River had cut through some small hill, could hardly call them canons being used to such large ones. In starting Fred broke one of his oars. Camped on the right bank in Swallow Cañon on a narrow ledge for dinner. The River through this Cañon is very quiet. Walls about from 500 feet to 800. Pulled out at 2 p.m. when we opened out again into the Valley. Ran down to a beautiful Cotton Wood Grove on the right bank where we remained to jerk our beef.

June 12, 1871

June 12. In camp all day mending and washing. Major, Steward and Jones went accross the River to get Geology and Topography. Beaman took picture of camp.

June 13, 1871

June 13. Pulled out early this morning. About a mile down we met Mr. Bacon and a Georgean Negroe. Mr. Bacon is a Brother-in-Law of Mr. Harroll. We had promised to ferry him accross to some cattle he had on the other side of the river. The water being up high he did not like to swim his horses, and therefore did not go over. Bid him goodbye and pulled out. Lashed boats together. River being very wide and deep we floated down while the Major read aloud the Lady of the Lake. On the left came in a Stream called the Vermilion, so called by Fremont. A stream of about 75 yard in width it starts about the Park, at the present it contains but little water. Camped on the left bank for dinner. After dinner dropped down to the head of the Canon of Lodore where we camped in a Box elder, or Aconigundo Grove. Musquitoes were so thick in the bottom that we had to move upon a bluff if we desired to keep all our blood. River still rising, bully for us.

June 14. In camp. Fixed the Major's Moccasins. Washed my clothing. Our soap is getting short. Fred and Steward went out this morning with two day ration, to geologise and sketch some distant mountains. Waiting here for Mr. Harroll to return from Green River with mail which had been sent to Salt Lake after our departure from Green River, for which he will telegraph to Mrs. Powell, who collects all of our mail. After dinner climbed mountain on left of Lodore, Prof., Major, Clem and myself.

June 15. Still waiting. I planted our flag on a high bluff overlooking the Valley below, as a guide for him to our camp, wondering if ever our flag had kissed the breeze from these lofty mountains. My friend Fred and Steward returned about 1 p.m. tired and hungry, having cached their provision the first day, but being unable to find them again had to do without until they made camp.

June 16, 1871

June 16. Still waiting in the evening our old friend Mr. Bacon rode into camp. Informed us that the mail carryer would be in late in the night or early next morning. He had supper with us but declined to share our hospitible ruff or that portion of the Canopy which covered our bluff.

June 17, 1871

June 17. Our mail carryer did not arrive till near noon when he brought only two letters, one for the Major, the other for Prof. The rest of our mail had been sent on to Uinta. Pulled out after dinner and entered the Rocky Gate of Lodore,[9] some Newspaper correspondent has said: he who enters here leaves all hope behind. Walls about 2500 feet, as we entered we could hear the roar of rappids. Ran two in fine style, shipped but little water. Beaman took pictures of rappids while the Dean went on alone for a few miles farther down but soon stopped and camped on the right bank a little above a roaring rappid nice music to go to sleep by.

June 18, 1871

June 18. Beaman took couple of negatives [of] a Grotto in the cliff. Steward called it Winnies Grotto.[10] Clear cool water running down the walls. Pulled out and ran that roaring rappid whose noise had lulled us to sleep. Done it handsomely. Shipped but little water. We watched the others come through as they had done from the shore. To see one of the boats run a rappid they resemble the bounding deer through a forest of fallen trees. Ran another bad rappid. Boats nearly filling,

[9] The name "Lodore" was apparently suggested by Robert Southey's poem, "The Cataract of Lodore," which Powell evidently knew by heart.

[10] Named for J. F. Steward's daughter (Walter Clement Powell, "Journal of W. C. Powell," ed. Charles Kelly, *Utah Historical Quarterly* 16–17 [1948–49]: 273; Dellenbaugh, *A Canyon Voyage*, p. 35).

plopping through the waves. Pulled hard to save ourselves from going over another before barking boats. Camped on the right bank for dinner. Ran two more rappids and then camped for the night ahead of the Falls where two years ago the "No Name" was lost. Major calls it Disaster Falls. Beaman narrowly escaped going over. All hands stood aghast for a moment, five feet more and he would have shared the fate of the No Name.

June 19. Ported our provision over a rocky hill, then let the boats down by line a distance of $\frac{1}{8}$ of a mile. Loaded up and let them down to the next fall. At this fall we found a sack of flour which had been rescued from the No Name two years before. Andy baked biscuits for dinner out of the flour. Found it in perfect condition with the exception of a crust of an inch thick on the outside of the flour. Unloaded boats again. Ported provision for $\frac{1}{2}$ mile over a hill rocky all the way. Let one boat down by line after lifting it over rocks. Men being completely tuckered out Major concluded to leave the others till morning.

June 20. Let the other two boats down. Loaded up and pulled out. Let the boats down by line around lower Disaster Falls. Had dinner among a lot of sage and cedar, made a short portage. Ran out quite a ways. Stopped just above Cascade Creek. Pulled out immediately after. Ran a bad rappid. Ran a little to close on an island. Got five very bad raps on our keel. Camped opposite a cliff 2800 feet high, called the Cliff of the Harp.[11] From this Camp a very peculiar mountain is visible. It is in the shape of a wheat stack, and therefore called so.

[11] Named by J. W. Powell on the second trip for the constellation Lyra (J. W. Powell, "Colorado River Exploration, 1871–1872," p. 15).

June 19, 1871

June 20, 1871

June 21, 1871

June 21. Let the boats down by line over a rapid found a vice and axe lost by party 2 years ago. Camped for dinner on the right bank, close to the waters edge. Ran a rushing rapid immediately after dinner. Two feet more to the right of the course we were running and the Emma Dean would have been shivered to piece[s on] a large rock projected from the wall. The chanel lead close passed it and the suction carryed us closer than desired. Ran down a few miles of rapids until we came to one which stopped our rushing speed. Landed and let boats down by line. Pulled out for half a mile, stopped before another "Rusher." Let down by line. Had hard work to clear away an old Cotton Wood Tree which obstructed our passage. When about to leave Fred missed his sketches and found that he had left them at the second Portage so he had to climb a steep cliff to get to them. We dropped down a little way where we waited his return and then pulled down to Head of Tripplet [Triplet] Falls.

June 22, 1871

June 22. In camp Major, Prof. and Cap climbed Dunns Cliff,[12] part of the Sierra Escalanta,[13] 2800. Washed my clothes and caught a mess of fish.

June 23, 1871

June 23. Unloaded the boats and let them down by line, ported provision a short distance over the rocks. Loaded again and let them down by line about a quarter of a mile, then unloaded, let the boats through a rapid and ported provisions some distance over a hill. Had dinner at the end of the portage and pulled out. Ran four rapids when we came to one whose noise we had heard for some time

[12] Named for William H. Dunn, who, with the Howland brothers, left the 1869 river party in the Grand Canyon, climbed out onto the north rim, and was slain by Shivwits Indians (see William C. Darrah, *Powell of the Colorado* [Princeton, 1951], pp. 140–41).

[13] Thompson, "Diary," p. 20, calls this "Serratus Ridge."

above the roar of all the others. Landed at the head of it, unloaded the Dean and let her down. As she was going through the shoot she struck on a rock and keeled over. Pulled her up, righted her and found but little water in her forward cabin, no damage done. Let her down some ways, ported provisions, loaded her and then took her down the stream by land the water was so shallow near the shore that we had to slide her along on the bottom for a half of a mile. The river is filled with large rocks with a fall of some twenty feet. The water rushes through these makes it one sheet of foam. We let the boats down for some distance when we unloaded her again and let her down for ¼ mile by line, ported provisions, climbing up a steep bank 50 feet high. Reloaded and then got in to pull her some distance to a little sandy beach. The quickest ride I have ever had in a boat I had while going this short distance. Made her fast and then returned to camp tired, but oh how hungry. Major very appropriately called it "Hells half mile".

June 24. Brought down the other two boats and got through just in time for dinner on the beach. The sage and cedar at our breakfast camp caught fire and the dense smoke rolled up to the peaks some 2800 feet in height and then out of "Lodore". After dinner pulled out, ran three tolerable rapids, and camped at the head of another. The Nelly in coming over one she ran on a rock, poised amid air and for a moment was doubtful of her fate, but a wave came and took her off nearly capsizing.

June 25. Fred, Steward, and I helped Beaman and Clem carry their instruments up the mountain from which Leaping Brook comes hopping down to the river. I attempted to sketch some of the falls, but failed to do justice to its beauty so gave it up in disgust. After dinner let boats down by line. Had to lift them over

June 24, 1871

June 25, 1871

the rocks and then again some ways by line. Got in and ran some bad rapids. The Dean shipped her standing room half full, the Canonita ran her bow on a rock but the current took her stern around and so slid off without damage. Reached Alcove Park about 4 p.m. Beaman took some pictures of a Creek of that name and its surrounding beauties. At this place two years ago the last Expedition after having been landed a short time, their camp caught fire, in which they lost most of their mess kit and some clothing. They had to "Git" to save themselves. Ran down to the mouth of Bear or Yampa River where we arrived at seven o'clock.[14] Major says will remain some days.

June 26, 1871

June 26. In camp fixing up for a trip up the River tomorrow. Loaded the boat with three days provision. Caught some fish.

June 27, 1871

June 27. Major, Jones, Beaman, Hatten and myself started out up the Yampa, Hatten and myself on the Oars. Made good headway for a few miles when the Current began to tell on us, until finally we could make no headway. We got out with our line and imitated the mules on the Towpath. Camped on the left bank for our dinner. Beaman took pictures. After dinner we had something more than strong currents to contend against for we heard the noise of a small rapid "around the corner". All hands got hold of the rope, Hatten and myself leading some ways. The shore being thickly fringed with willows made it an awkward thing for us to get her over a fall of about 3 feet. Major proposed to get her over stern first. Hatten and myself went back to the boat to help lift her over some rocks. Got her over all right, Major, Jones and Beaman holding on while Hatten and myself went to the

[14] Powell's men called this area Echo Park (Jones, "Journal," p. 42; Dellenbaugh, *A Canyon Voyage*, p. 49); it is also called Pat's Hole.

end of the line to hold on while they came to our assistance. They had hardly left the boat when the boat bulged and down she went but fortunately was stopped by Hatten. The rope had got around his leg which was fast between the rocks and myself holding on like grim death to a negroe, we stopped her half way over. The rest of the party coming to the rescue soon brought her up when a second time she got away but stopped herself by bumping against a rock. Fortunately for Hatten he received no injuries. Got in and rowed a little ways, then had to get out again to tow it, so by alternate rowing and towing we got up some 4⅛. Camped on the right bank. Beaman tried to take a picture of Canon with moon in but failed to get the moon. Walls from 500 to 1000, mostly lime and sandstone. Major called this Yampa Canon and the little park Grisley Park on account of the numerous tracks we saw of bruin, but failed to see his Majesty himself.

June 28. Major and Jones climbed out but failed to get high enough for Topographical observations. Had dinner. So we harnessed ourselves up again and towed and pulled up a rappid, after passing this we got in and pulled for some distance. Saw some mountain sheep, tried our guns but failed to hit. Little farther up we saw seven high up on a hill looking down on us. Camped on the right bank on a narrow ledge under a box elder. Major and myself had our bed to slanting and so kept sliding down the "Cellar door" all night.

June 29. Pulled out early, ran up a short distance and then got out to tow. Camped on the right bank about 10, had a lunch. Major climbed out but failed to get high enough. Returned and reported some fine views from the top of mountain, so we took Beaman's instruments and all started up. Returned about 1 p.m. Started out with the tow line and whenever we had a chance we pulled. Major shot

June 28, 1871

June 29, 1871

39

at some geese but failed to hit. We tried but followed his example. Then we concluded that they were "Wormy, Story of Fox and Grapes." About four p.m. we landed on the left bank opposite an island. River taking a sharp bend to the left looking up three latteral Canon[s] coming in here, in one of which is a creek of pure sparkling water. The "General" Hatten went up and explored it. A little farther up a beautiful park opens out. Major called it after him and so it goes down on the map as Hatten's Park and Hattens Creek. Found our rations short, only enough for breakfast. I caught a mess of fish in lieu of bacon.

June 30. Passed a very cold night, the coldest on the trip. Major "spooned" up like a good fellow. Got up very early. Helped Hatten get breakfast ready. At sunrise Major and Jones started to climb a bald mountain some two miles from camp. Beaman, Hatten and myself dropped down stream to a latteral Canon for some picture on the right. Stopped at the head of a rapid, fastened the boat, carried the instruments some distance down the stream. Hatten and myself returned to the boat, rowed up some ways and then crossed to the left bank to wait for Jones and Major and to prepare dinner, but I had to catch it out of the river first. No coffee or tea, no bacon. All we had was one biscuit and some sugar. I was lucky enough to catch five large fish, one apiece, which we roasted. About one o'clock the Major and Jones returned, having been very successful in getting observations. I gave them one biscuit and one fish apiece and all the water they wanted to drink — but Jones was sick of fish, he begged me to take them out of sight. This finished, we dropped down to where Beaman was waiting for us. Issued him his ration of one biscuit and fish which he ate while Hatten and myself pulled for our supper on the Green as best we knew how. Made three portages by line and lifted boat over rocks twice. Made all told 12 miles. Found Clem and my friend Fred in camp in charge of culinary department, the rest having climbed a mountain, who soon after returned and that evening was spent in telling adventures.

July 1. Rested and cleaned up.

July 2. Made Major a [pair] of moccasins and mended my shoes.

July 3. Started this morning from Echo Park as the Major called it on account of the great echo, there being a perfect wall of solid sandstone some 600 feet in height it runs some 500 yard along the river at a thickness of 200 feet at the base and quite sharp on top [Steamboat Rock]. The River turns back on the other side, making a sharp bend round the point of the wall. Left the Canonita behind. Beaman wanted to take some pictures immediately after making this bend we entered Whirlpool Canon. At the entrance we were saluted by the roar of a rapid which we ran. Walls 500 feet and sloping from the river to 2500 feet in hight. Passed over two rapids, then came to one which we thought best not to run. Waited for the Canonita but did not come in sight until we had the Dean and Nell let down by line. Had dinner on the right bank opposite Rapid. Let Canonita down and started. While at Echo Park water fell 3 feet which helped us considerable going through this Canon. The river is not so swift and therefore less whirlpools. The whirlpools are occassioned by the rush of water against the ragged rocks which project from the wall. Nearly the whole wall presents the appearance of a stair laid down on its side. Saw some 9 or 10 sheep on the right bank high up on a Tailess [talus]. Jones and Fred started up to cut them off but they were too late. Camped on the right bank at the mouth of Brush Creek,[15] so named by Fremont, a nice little stream of pure water, full of trout. I started up for a mess but they were too stubborn to take my grasshopper.

July 4, 1871

July 4. Oh glorious Fourth. The boys saluted the sun with one shot for every state in the Union. Major gave us a Holyday, but Prof, Jones and Hattan climbed out and did not return until night. I being almost barefooted thought of a plan to make myself a pair of "Wooden Shoes". I succeeded. Made the soles and sides of wood and covered the top with canvas. I congratulated myself, but low, while walking over the rocks the bottom of one split, so my whole days work had been in vain. At 6 p.m. we had dinner, but such a dinner not to be sneesed at in this wild country. I had procured some canned fruit while at Green River, which I had stowed away in the forward cabin so that no one knew except Fred, he had a few pounds of candy which he had brought, Hattan being away so Fred cooked dinner. Had a lunch about 12 n. No one suspected what we had for dinner, everybody being surprised to find strawberries, peaches, tomatoes, pies and candies, ham and a beautiful strong cup of tea. Everybody ate hearty and enjoyed it more and better than they would have done at home. After dinner lit our pipes and cursed the Irish until it was time to seek the fond embrace of Morpheus, and so ended the glorious Fourth of July of 1871.

July 5, 1871

July 5. Pulled out this morning. Ran some half dozen rapids, bumped on rocks here and there but not serious. Camped on left bank under a cottonwood tree

---

[15] J. W. Powell later renamed this creek "Bishop's Creek" for F. M. Bishop; it is now called "Jones Hole Creek" for S. V. Jones (Jones, "Journal," p. 45, n. 24).

for dinner. Jones, Bishop and myself went across the river to shoot some sheep we saw on the hill but failed. After dinner pulled out, ran a few more rappids, at one place barely saved ourself from being crushed to pieces by rocks. We ran to[o] close not expecting such strong suction toward them. Ran one more, an old monster and then opened out into a park, the little low hills being highly colored and its different shades were as numerous as that of the Rainbow, in Horizontal stripes. Major called this Island Park. The river breaks up into small streams as it runs through this low country, forming numerous little Islands. The Park is about 8 miles long by River. Camped on the right bank in a Box Elder Grove, the Head of Craggy Canon [Split Mountain Canyon].

July 6. In camp, fishing and fixing up my traps. Major, Prof., and Steward went out for fossils, returned at noon brought in a few, went out again after dinner, and returned about 4 p.m. with quite a lot. Major asked me how far I could walk in a day and I was surprised. I answered carelessly. Oh about 30 miles. When he asked me if I could not make 40 I told him that I might at a pinch. Then he said that he had notion to go to the Uintah Agency where our rations are stored, it being 60 miles but we had to make 40 miles the first day in order to reach water. Talked it over with Prof. who persuaded him to take boat and go as far as the Uintah [Duchesne] from there foot it, so he finally concluded to go by boat. Made arrangements accordingly. Exchanged Fred for Capt. Bishop.

July 7. Got up early and pulled out at sunrise. Had not gone far before we struck a rapid which we ran, little farther down still another, made a portage by lifting the boat over the rocks, worked about 40 minutes, had hard work, pulled out a little ways and then let down by line. Rapids all the way, got in pulled across to the right bank, made a portage by line. Boat got stuck on a rock, Jones went in the water a little ways to push her. He gave her push when all of a sudden she slid off. Jones losing his hold went headlong in the River. All I could see of him was his white hat. Fortunately for him he got hold of a rock, thereby saving a swim down the rapid and perhaps his life, Having left the other two boats behind to finish up getting Topography and Geology. They will spend three days in this Canon for that work and also for Beaman to get pictures. After letting the boat down we got in to pull but soon were stopped again by rapid. River breaking up into two streams, forming an Island in the centre. Landed on this. Took the right hand chanel, let the boat down by line. At the foot of the Island got in for a short distance ran a rapid and camped on the right bank for dinner. Immediately after pulled out, ran a rapid, then led down by line. Pulled a little ways, then had to get out and let her down again by line. Ran two more, then came to one a huge old monster. Let down again by line. At this point we could see the end of the Canon. At the end we mistook the chanel and got into the wrong one. Had to get out to lift the boat over the rocks. Then rushed down the river at 10 knots an hour, could fairly see it go down hill. Passed over some small rapids, all o.k. Then we opened out into the Valley of the Uintah. Craggy canon is formed of a mountain split in two by the River. Major first called it split mountain, but changed it to Craggy. Innumerable crags, peaks and Pinnacles can be seen. Ridges running perpendicular to the River. The scenery is grand. Some peaks ran up to nearly 3000 feet. Those huge old crags bespeak age, a fine study for Geologists, but I decline to puzzle my brain with the age of a stone. Got out of the Canon about 4 p.m., having made seven miles, pulled in seemingly smooth water for seven miles more then camped on the left bank under some cottonwood, near a lot of Indian "Wickiups", Wigwams.

Ran 19 miles, killed three geese but on getting the feathers off them we found them to poor to eat.

July 8. Got up early and after breakfast pulled out, leaving our geese on the bank for wolves and coyottes. Here the river bends and twists through the Valley very wide and shallow. Had to get out to push boat over bars several times. Camped on the right bank for dinner. Immediately after dinner pulled out again. Rained and blew fearfully, mixed with peals of thunder. Camped on the right bank for the night. Shot some geese in the afternoon which I picked for the feathers to make myself a pillow. Found one to be pretty good. I cleaned it for breakfast. Plenty of mosquitos here, could hardly sleep for the rascals.

July 9. Got up early. Prepared my goose and some ham while Capt. Bishop baked the biscuits and Jones brought food for fire. I found my goose too tough for my ivories, so left it for the wild beasts. Pulled out while the Major read about Harrold the Dauntless, one of Scott's poems, hardly any current, had to pull her through in order to make headway. While passing near a little knoll we saw two antelopes, one drinking, the other on the hill coming down. Dropped down easy and then landed below the hill. Major got out with gun had a good sight for killing them. He cocked his gun, aimed, fired, but low it missed. The noise frightened them and away they bounded. On examination found the shell of an exploded cartridge in the Bore, in place of a primed one. Moral, before you shoot at wild game be sure that your gun is loaded or 10 to one you miss getting it. Passed Powells Lake on the right. Major mistook the outlet first for the mouth of the Uintah but soon found his mistake. This lake has been called after his Brother, Capt. Morris [*sic*, Walter H.] Powell who two years ago was with the Major on the former expedition. It is some

July 8, 1871

July 9, 1871

45

two miles wide and about 3 or 4 in length and contains very clear water. Camped on the left bank for dinner. Pulled out as soon as we had our chuck down. Ran until 4 p.m. when the Major pointed out a large cottonwood Tree as our destination, which we reached a little afterwards. Found a house on the left bank, pulled over and found it vacated [Figure 4]. Major thinks it belongs to a man by the name of [Pardyn] Dodds, the former agent for the Uinta Utes Indians. Pulled to the other side again where we camped for the night. The Uintah River joins the Green about ⅛ of mile farther down. At our Camp is the Old Overland Stage Companys Ferry. The Pioneers Wagon tracks are still visible. The boat which was intended as a Ferry boat is sunk here on the bank, she having been hauled close to shore and filled with sand to sink her, two stakes being driven in to keep her from going down stream in case of the water washing the sand out, but no danger of that. Where once was the edge of the River is now almost highland excepting high water mark. This road has never been used. Only one train went over it with the constructing party. It runs from Denver to Salt Lake in an almost straight line, almost from East to west.

July 10, 1871

July 10.[16] Got up very early, packed the Major's blankets and mine into a little knapsack, got a days ration into a bag, had breakfast about five. Major equiped himself with the ration and a Henry Rifle, while I being an old soldier slung knapsacks. About 6 we started out for the Uintah Agency. Made 12 miles in four hours. Where we crossed the Uintah River had a wash and a smoke.

[16] Hillers's diary is the only extant description of the hike made by him and J. W. Powell to the Uintah Indian agency. J. W. Powell's diary ("Colorado River Exploration, 1871–1872") stops on July 7, 1871, and resumes on September 2, 1871, when he rejoined the river party at Gunnison's Crossing.

Forded and struck out. Made some 8 miles more when we came to the River again. Here we concluded we would have a cup of coffee. I soon had it made. Drank and smoked, then up and off again until about 6 p.m. when we struck water again. Had a cup of coffee and smoke, when we crossed several branches. Walked until nine p.m. when we came to a crossing where we lost the trail in the woods. All our efforts to find it proved fruitless so we concluded to camp out and walk in early in the morning. Laid our few blankets on the ground and turned in. Slept sound.

July 11. got up early, cooked our coffee, smoked and started off soon finding our trail. Walked a half an hour when we saw the Agency at whose breakfast table we found ourselves soon after doing justice to a hearty meal. Found lots of mail for the boys, one for me from Dick. People have called it 45 miles but Major called it about 40. Found the Lord of the Manor gone, had started three days before our arrival to Salt Lake City. A Mr. [J. J.] Critchlow by name is the Agent for the Uintah Ute Indians, a sort of second hand Prespitarian Minister, and as I understand from some of his men, he is more Mulish than rightious. Before leaving he issued orders to his foreman that under no circumstances he was to let Major Powell have teams to carry our rations down. It took us aback. We did not know what to do, but Mr. Besser [George Basor] the Trader came to our rescue. He offered us his team. Now supposing no teams were to be had, would we pack our rations down 45 miles on our backs? I think not. I would have harnessed up one of his teams without permission and drove it down to Green River and then returned. I don't think one of his men would have resisted my taking the team. The men say when he took charge he wrote out regulations and stuck them up at the Carpinter and Blacksmiths Shop. The first thing he done was to abolish the ten hour system, men had to work from sunrise to sundown, one hour for dinner. All Government work according to the latest law is only for 8 hours during the day, extra pay being paid for overtime. On the whole I dont think much of Critchlow taking the word of his men, and his regulations confirmes them. Found eight buildings and lumber for another. One office, one Blacksmiths and Carpinter shop, one for mens quarters, one for kitchen, one for stores, and one barn. Mr. Besser and Mr. Dodds each own a house. Mr. Dodds has a large drove of cattle in this Valley, but talks of going farther down the River, His Royal Highness, the King of the Valley having hinted that he must take his stock away, that the Valley belonged to the Indian Reservation. Good many [of] the Indians came to pay [the] "Americats" a visit expecting of course heaps of "shug" and "frour" in return. Smoked the Pipe of Peace with Old Tuckanoana,[17] a fine looking man I should think about 45 *snows* old and 6 feet in his stockings. His father used to be chief of this tribe, but he declined, he went to farming, the only one who seems to go at it with a will. He has got a nice little farm fenced in. Talbia,[18] a fine looking man, is chief at present about the same age, married or rather took unto himself a nice looking Squaw of about 16 snows, the best featured woman I have seen among the whole tribe. The Squaws do all the work while the head of the family takes it cool by picking the lice off himself which abound in his dirty carcass. Whatever they put on it stays there until it rots off. They never think of taking it off to wash. Their breeches are of Buckskin and are sewn on their legs skin tight, with sinnew or "Tammu" as they call it. The women look dirty, some wear Buckskin dresses, fixed with beads. A dress of this kind is valued at about 50 dollars. They carry their children on a piece of bord with a basket on top as a cover for rain or sunshine. The "Pappoose" is tied to the bord and

[17] Jones, "Journal," p. 49, calls this man "To-a-quan-av."

[18] This man was usually known as "Tabby" (see K. B. Carter, comp., *Indian Chiefs of Pioneer Days*, 2nd ed. [Salt Lake City, 1932]).

slung on the Squaws back. It is very seldom that you heard one cry, and they are called pappooses until they get married. Major took possession of Mr. Critchlows Office, in which we slept.

July 12. Had a good nights rest. Got up, had breakfast at an early hour. Major talked of going to Salt Lake City. Mr. Dodds offered his horses and thought of going himself. Had a look at our rations, found them in good order, but no salt or soap, articles we were very much in need of, good many other articles which had not been sent from Camp Douglass, U.[tah] T[erritory]. The most I missed myself was ham. The Major had bought and paid for these things and why they did not send them is more than I can tell. Made an exchange of bacon and sugar with Mr. Laighton, [Thomas Layton] the man in charge for some soap enough to last us for two months. Also took half of a hide for soles of moccasins. More Indians today, gave them a sack of flour to devide among themselves. How they did it I cannot tell, but I think it remained with the two aforenamed Indians to whom it had been given for distribution. Major concluded to go to Salt Lake City tomorrow morning, leaving word and letters for Mr. Thompson how to act. Got everything ready for his journey tomorrow.

July 13. Major and Mr. Dodds left early this morning. After breakfast I went down in the meadow for the horses, wanting to get Mr. Camble's [J. L. Campbell] horse shod for me to ride down with to the boats, and let Mr. Thompson ride up who was expected to be at the crossing about this time. As I got back with the horses I met Capt. and Jones. I was glad to see them. They had come up the Uintah some distance with the Boat and there cached it under some willows. They were a tired pair when they got here. They too had lost the trail and camped on the same ground on which the Major and myself had camped on before. Shoeing Camble's Horse, I ran her for some five miles as hard as she could run, to take some of the fire out of him, which made my thies very sore not being used to riding. Spent the rest of the day talking and resting.

July 14, 1871

July 14. Cap, Jones and Mr. Layton went down to the Indian farm some four miles from here. They returned at noon. I made a pair of moccasins for Mr. Black, one of the farm hands, in whom I recognized an Old Brother Soldier having served in one Brigade together, with Butler in front of Richmond. Had quite a chat about Old Times Rocks. Three Indians belonging to the White River Reservation came riding past the Agency about 9 p.m. I had just "turned in" near a Haystack when I heard three shots fired. I thought some of the boys were discharging their pieces though I heard the whiz of the bullet when immediately after I heard Jones call me. I answered but did not get up, thinking that he was coming to bed and wanted to know if I had retired, when a second time he yeld like a stuck bull for me to come up. I went up to where I saw them all standing, was surprised to find them all armed. Of course I got my Henry XVII and joined them to slay anything that might [be] hostile to my life or good for my stomach. Mr. Besser the Trader soon joined us and said that the Indians who had fired had asked him if he was friendly to Indians. When answering in the affirmative they rode on. Little while after we heard the War whoop, which sounded like business. We remained on our guard until 10:30 when we turned in after barring everything.[19]

[19] Following the incident, " 'Bismark' and I concluded to move our bed within doors and slept for the first and only night since May 20th under a roof" (Jones, "Journal," p. 50).

July 15. Heard this morning that the three Indians who fired were excited about a murder that had been committed near White River, and 10 had died from some kind of epedimic, the murderer had escaped, belonging to the same tribe. Capt. and Jones left this morning saving me a ride down to the boats on horseback. Fixed my shoes and made myself a pair of moccasins

July 16. Helped to unload hay to keep myself busy. Expected Prof. and Beaman up but did not come.

July 17. Helped the Blacksmith a Mr. [Martin] Morgan from San Francisco to weld out irons for the keels of our boats, which needed them very bad. A mistake of the Major, thinking that it would make them too heavy to have the keel ironed. In going over the rocks the wood was almost worn even with the planks so it was necessary to have irons put on. No Prof. this day.

July 18. Finished the irons and read Scribbner's Magazine. Prof. and Beaman came up about 7 p.m.

July 19. Got our rations ready for tommorrows start. Helped Mr. Besser fix up his wagon and got my own traps ready.

July 20. Had breakfast, loaded our rations, got started about 8 a.m. leaving Prof. and Beaman behind to photograph some of the Indians and the Agency, or such views as they might find of worth. Besser and myself started slowly, day being very warm and no breeze. Camped for the night at the Uintah crossing. Fastned two horses with ropes while we hobbled the other horse and the mule, having crossed before we camped, thought them secure. About 12 at night I was awakened by plunges in the River. Thinking that they had gone down to drink waited for some time, but they did not return, so I concluded they had crossed, but could hardly satisfy myself about the horse and mule crossing they being hobbled, and how they could cross a rapid was more than I could account for, but I could see no trace of them so I crossed. Ran through the woods when I laid down to listen, but could not hear them. I ran half mile farther when I listned again, when I thought I heard something bound. I started and found them making their way back. I got hold of one rope, mounted him and drove the rest back. They had gone two miles. I had a fine shirt tail Parade, but no one to review me. Had them picketed afterward.

July 21, 1871

July 21. Started out at sunrise. After going up one hill and down another we at last reached Green River about noon. Shouted for the boys who soon made their appearance on the other side. All hands came over to welcome me back an indeed I was glad to be "at Home" with the boys. Mr. Besser had dinner with us and started back about 2 p.m. Rested the rest of the day.

July 22, 1871

July 22. Washed my clothing and done a little mending.

July 23, 1871

July 23. Made the Major a pair of moccasins and soled his old pair.

July 24, 1871

July 24. Having made these so well, everybody wanted a pair, and being all barefooted I made a pair for the whole outfit. Commenced and finished Capt's. Fred, Steward and Cap started up White River took three days rations.

July 25. Saw some antelope this morning. Started out for them but they saw us before we got within range.

July 26. Took the Dean out of the water, put iron on her keel and caulked and pitched her.

July 27. Finished the last pair of moccasins. Sawed slats for cabins. Fred Steward and Cap returned at night having gone up 50 miles both floated down on a raft.

July 28. Washed and fished.

July 29. Went out hunting but could not see a single game. Prof., Beaman and an Indian returned from Uintah. They had horses from the Agency which the Indian was to take back. Major had gone to the mouth of Dirty Devil [River].

July 30. The Indian started back this morning.

July 31. Commenced packing our rations in the boats. An Indian made his appearance on the opposite bank. Went over and found a son of nature and his Squaw. He proved to be the son of Douglass, the Chief of the White River Indians. This young hero was on an elopement tour, having run off with another chiefs daughter, she being another bucks promised bride. These two showed open affection for each other, the only instance of the kind I have ever noticed among them.

July 25, 1871

July 26, 1871

July 27, 1871

July 28, 1871

July 29, 1871

July 30, 1871

July 31, 1871

August 1, 1871

August 1. Beaman took a photograph of these two children of nature. Had no objection to have their picture taken.

August 2, 1871

August 2. Prof took latitude and time while the rest worked at the boats.

August 3, 1871

August 3. Our Indians paid us another visit. The Squaw made Jones a pair of moccasins but did not make them to fit good. I went over where she made them but dropped her work as I approached. Asked if she had enough of buckskin taken she said she had. I thought she had taken enough for two pair so I overhauled her storage and found enough stuff to make herself a pair, but I took it out of her bag. She only laughed. These Indians will steal and when caught only laugh.

August 4, 1871

August 4. Made everything ready for tommorrows start. Our Indians bid us goodby. He started down the River. We christned him Louchinvar.

August 5, 1871

August 5. The long looked for day had come at last and at 8 a.m. we pushed off. Nell leading, Dean next, and the Canonita bringing up the rear. Passed the Uintah about 50 yards wide on the right, the proper name for the Uintah here is DuSaine [Duchesne] or at least the Indians call it so. The Uintah joins the DuSaine about 12 miles from its mouth, a stream not half the size of the DuSaine, and why the latter should lose its name I cannot tell. The DuSaine starts at the Wasatch Mountains and the Uintah from the mountains from which it has its name. The two combined supply the Green with considerable water. About a mile below on the left we passed the "White", a stream about the size of the Uintah. It starts from the Parks in Colorado. Made 10 miles in the forenoon. Beaman killed a beaver but failed to get him, he sunk. A little farther down we met our Old Friend Louchinvar and his other greasy half who was jerking some venison. We exchanged a few pounds of "shug" for some Jerk and pulled out for a few miles. Then camped on the right for dinner and observation. Pulled out again at 2:15. River very quiet and full of sand bars. Got out several times to push boat over. Ran 9 miles this afternoon. Camped on an Island. Called it Beaver Island. Called it Beaver Island on account of the many here.

August 6. Concluded to stay in camp this being Sunday. Prof., Steward, Jones and myself climbed out on the left. We left the valley about 3 p.m. yesterday and are gradually running into what is known as the Little Mountains by the people of this country, but not laid down on the map. Looking from the top it is the most dessolate looking country I have ever seen. Even the hardy sage dies for the want of nurishment. The only sign of life is on the bank of the River wherever it is sandy the willow and cottonwood grow.

August 7. Pulled out this morning early. River still very quiet and full of bars. Got out and in most of the time to push boat over. Shot a beaver but failed to get him. He slipped out of my hands and sunk. Camped on the right bank for dinner. Prof., Steward and Cap climbed out some 1050 feet for observation. Started at 3 p.m. Dropped slowly down. About five o'clock Prof. saw a beaver on the bank which he killed, rolled down to the waters edge where he wollowed in the mud. Steward jumped out but before he got him into the boat he was covered with mud from head to foot. Camped on an island. Bish and Clem skinned him and so in place of oxtail we will have beavertail soup for dinner tommorrow. Passed a weep-ing willow tree on the right, first I have seen on the trip. Made 13 and 3/5 miles this day.

August 8, 1871

August 8. Had beaver steak for breakfast. Pulled out at 8 a.m. Ran a few miles. Camped on the left bank. Boys got sick of beaver so in place of beavertail soup we had bean soup for dinner. Beaman got some fine pictures of a latteral Cañon. Pulled out after dinner for about a mile and a half. Beaman saw some nice views so we stopped. Ran a rapid about a half mile after leaving dinner camp which we called the commencement of the Cañon of Dessolation. Camped on the left bank for the night. 3 and ⅜ miles.

August 9, 1871

August 9. Camp this day. Prof, Beaman, Clem, Fred and myself climbed a small mountain for to get a photograph of an amphitheatre, but failed to get it on account of the darkness of the day. Left the instruments on the mountain for to-morrow. Spent the rest of the day washing my clothes. Walls opposite camp 800.

August 10, 1871

August 10. Prof, Beaman, Jones, Clem and Fred started up the mountain on which we had left the instruments. Being a fine day they got a beautiful picture. The river doubles up at this point and runs back within 200 yards of itself, for nearly a mile a narrow wall dividing it. Party returned at 11 a.m. Cap. and myself cleaned and filled barometer. Had dinner and pulled out about 2 p.m. Stopped for observation on the right. Beaman took some pictures of an alcove. Heard the noise of a rapid once more which we ran. The Dean struck a rock in going over but no damage. Camped on the left bank behind a line of Box Elders. Got out once this day to push boat over sand bars. Made 10 miles this afternoon.

[Aug.] 11. [1871] Prof, Bishop and Steward climbed out this morning for observation. Returned about dinner, while we dropped down about an ⅛ of mile. Andy and myself got dinner while Beaman took pictures. After dinner pulled out, ran down for about 3 miles when we struck a nest of rapids. Ran the first one in fine style. The second looked better than the first, but as the Nell went over, struck a rock, but slid off. Jones, Fred and myself with the Dean struck the same rock, but being a little heavier loaded, our stern hung on. We pushed a while but to no purpose, so I jumped out at the stern, obtained a foothold on the rock, lifted her up, and off she slid. The Canonita struck a rock about the same time farther up which brought her broadside to. In swinging around she struck another which broke two ribs, leaked very bad, and had to haul her on the beach to brace her side. Delay one hour and forty minutes. Just below us was another which we ran all OK, making three rapids in one-half mile. Ran two more. Struck rocks in both, being very shallow. Got out to ease her load. She went over all right. Camped on the left bank in a cotton wood grove. Afternoon work five rapids and made 6 miles. Boys shot at some beavers but failed to hit.

[Aug.] 12. Started out this morning early with a rapid immediately after starting. All the boats stranded on it. Got out and pushed them over, being very shallow. Ran along for a mile when we came to a lot of rapids. Ran them all. Had to get out in two to push boat over. Saw a deer drinking. Landed and tried to cut off his retreat but he escaped by a gulch. Ran 8 rapids and made 5-⅞ miles. Camped on the left bank in a cotton wood grove for dinner. Concluded to stay all day. Fred and myself climbed out for some pine pitch. We had a hard time getting up, being very steep and shaly. While climbing along a tailess I got hold of a rock for support which gave way but I steadied it against my leg while I, with my other hand, got hold of shrub and thereby saved myself a descent of some 500 feet. Procured quite a lot of pitch. Our descent was anything but pleasant, with sliding and climbing we at last reached the bottom. Returned to camp just as the boys sat down to supper.

The walls of the canon range from 900 to 3000 feet, sloping backward, very ragged and craggy. The wear and tare of ages have formed innumerable ridges which run at regular intervals from top to bottom. The gulch between two ridges resembles the Devil's Staircase. The mountains here are covered with red pine with a mixture of pinions.

August 13, 1871

[Aug.] 13. In camp, being Sunday. Prof., Steward and Jones climbed out for observations. Returned about one o'clock. Rest of the day spent in resting.

August 14, 1871

[Aug.] 14. Started 7:30. Immediately after starting had to get out in a rapid and guide boats through rocks. Little farther down ran through a narrow channel. The Nell let down by line while Fred and myself held on to the side of the Dean and got through all right. Ran three more when we came to a very bad one, the water fairly boiling over the rocks. Concluded to make line portage, Fastened a line to her stern and bow to hold her with while two men guided her through the rocks. Being very deep in some places they had to hold on to the sides and float along with her. Got them over all right. Had dinner on the right bank under a cotton wood tree. Beaman took pictures of walls, which Prof. called "Fretting Water Falls." Fell some six feet. After dinner ran a bad rapid of ¼ mile long. It fell some ten feet. In one place could fairly feel the boat sink down. This was the commencement of the rapid. It fell abruptly three feet. It was swift riding all the way, the boats bounding like a deer. Got through all right. The Nell and Canonita both struck rock but not damaged. Landed at the head of another. Let boats down by line. Camped on the left bank under some cotton wood. Just before getting into camp had to get out and push boat over a riffle. Ran five and seven eights of a mile.

[Aug.] 15. Started out 10 AM. The reason for not starting sooner was on account of Beaman taking pictures. Ran down a quarter of a mile when we came to a fall, Called it Five Point Fall, on account of five mountains bring[ing] their ridges together here. Let boats down by line one at a time. I barked my shin very bad while guiding the Dean through. While letting the Nell down the Deacon and myself had hold of the rope when, in the middle, he slipped and fell, pulling me in with him. The consequences was a ducking. We looked like two wet bulb thermometers. Got over all right when we came to another which we had to run, but struck a rock. Fred and myself got out and guided her most the way ¼ mile long. Got in, ran half mile farther, where we camped at the head of another rapid for dinner. Started out immediately after by letting the boats down by line. Got out in about an hour and a half. Pulled for half a mile, when we came to another fall. Let down by line. Took us rest of afternoon. Camped on the right bank. Made 3-⅜ miles. Mountains very irregular, being all cut up with lateral canon, the reason for so many rapids, the debris being washed and tumbled into the river.

[Aug.] 16. Remained in camp. Caulked boat and pitched her seams. Prof. and Jones went across the river in order to climb a mountain but returned in about an hour after starting, not being able to see far enough on account of the hazie atmosphere.

[Aug.] 17. Prof., Jones and Steward climbed out for observation, while Fred and myself climbed out for pine gum, of which we found some six pounds mostly from pinions. Returned about 10:30, while the others did not get back till about 1:30 PM. After they had eaten their dinner we started out. Had not gone over a half mile when we struck a rapid which we ran—the Nell and the Canonita ran over all right while the Dean struck and hung on. Fred and myself got out and shoved her over. About one-quarter mile farther down we came to a fall of some seven feet. Let down by line while four men guided her through the rocks, sometimes walking on rock while others had to hang on to the boat for dear life. Made the portage in thirty minutes. The party two years ago ran this but got swamped and lost their barometers and three rifles. The former were recovered by a long search. Pulled out, but could hear the roar of another fall ahead, which we found about a half mile farther down. Landed and let down by line. We had hard work getting the boats over. Water very swift and fall some eight feet. Got over all right. About a quarter of mile farther we had a rapid which we ran but got hung. Got out and pushed boats over, got in to pull. Landed at the mouth of Nine Mile Creek just at the head of another bad rapid. I walked up the creek for some ways. It is a nice little stream of pure water, started by springs. Who ever gave it that name I do not know. We let down by line. The Nell tried it without line but got hung when the current caught her bow and swung her round like a top, striking on a rock at the same time, which nearly turned her over. A little farther down she struck heavy starting her planking. Camped on the right bank. Opposite our camp is the most highest vertical cliff we have seen on the river. It is some 2800 feet high. On the top is a very singular rock. It resembles the cabins of this country. We called it [Log] Cabin Cliff. Mr. Beaman photographed it. The mountains through which we are passing at present are not laid down on the map as such, but as a platteaux, but by old hunters called the Little Mountains.

[Aug.] 18. Mr. Beaman, Clem and myself climbed a small mountain for the purpose of getting a view of Nine Mile Creek. Took four pictures and started back for camp. Got in about 10. Packed our traps in the boat and started out. A quarter of a mile from camp we ran a rapid all ok. A quarter farther down we ran a bad rapid of ⅛ [mile] in length. Water very swift. Fell some 8 feet. Landed at the head of another to see where it could run. Found an appearing [illegible] channel, but the Dean, in running over got it a little too far to the left, and the consequence was that we struck a rock. Fred and myself jumped out on the rocks and shoved her over. No damage. Little farther down we heard the roar of a fall. Landed at the head. Took cooking utencils and while Andy prepared dinner we let the boats down by line. All safe. Came back, had dinner, carryed the cook's thing down and pulled out. Noon pulled out and ran a bad rapid. Little farther still another. Landed at the head of a fall, let down by line. Called it Melon Falls. Water fell some nine feet. Pulled out for half mile. Stopped to examine a rapid, which we ran. Full of rocks. The Nell struck but no damage. Landed at the foot of a rapid on left bank where we camped for the night. Made ⅝ mile.

[Aug.] 19. Pulled out this morning with rapid in sight, which we ran. In going over the Dean struck but slid off all right. Landed at the head of another. Found it an easy task to run over if we struck it in the right place. Happily, we did and went whooping for ¼ of a mile. ⅛ farther we landed at the head of a fall, let down by line while the Canonita came over. She got away from Andy, that is, she swung her bow round. The current took her round in a jiffy, pulled Andy to the

Ground but sustained no injury. Pulled out for half mile to the head of another rapid. Concluded to run it, which we did in good order. Half mile farther down landed on the left bank at the head of a rapid, went over to the right bank and let down by line. Got over all right. Pulled out and ran for a mile and a half in smooth water, when we landed on the right bank at the head of a fall. Let down by line, pulled out, and landed at the head of an old stunner. Had dinner and then let down the Canonita for Beaman to take pictures of the boats, as they came over the falls. Then came the Nell and the Dean last. At this fall a brook comes in on the left. Clear, sparkling water comes down over the rocks, boiling. It is started by some springs. Called it Chandler Fall and Chandler Creek. Fall of water 12 feet. Pulled out leaving the Canonita behind for Beaman to fix up his pictures. Ran but a little ways when we landed on the left bank. The Canonita soon hove in sight but landed to get some views of the brook, while the Nell and the Dean pulled out for a short distance. Landed on the right bank at the head of a rapid, where we camped for the night. Prof., Steward and Jones climbed out, returned at dark. The Canonita came in a little before.

August 20, 1871

[Aug.] 20. Sunday. Pulled out this morning, not being able to get a rest here on account of the many ants. Crossed over and let down the rapid by line. At the foot was still another, which we concluded to run. Got over all right, though it was chance work. Canonita stayed behind for Beaman to get a photograph of a natural bridge which spanned a gulch some 200 feet wide and 300 ft. deep, about 1500 ft. from the river. This bridge has been formed by erosion. Camped on the right bank under cotton woods. Traces of Indian encampment could be found in the shape of wickiup poles and brands of wood. I spent the day fixing moccasins. 1-¼ mile.

[Aug.] 21. Pulled out this morning for half mile when we came to a rapid. Ran in and jumped out in the centre, where we struck. Pushed her over and landed at the head of a fall. The Canonita ran over all right. Let down by line and pulled out to run another. Got in but little ways when we had to get out and push boat over, but soon after struck a channel. Jumped in rode at the rate of 20 miles an hour for an 1/8, went on for 3/8 (?) more in quiet water. Landed at the head of a rapid on the left bank. Walls of the canon running down. Distance between walls about 3/4 mile. I think by the appearance we will soon be out of this Canon of Desolation [Figure 5]. Let down by line, jumped in and pulled across to the right bank. Let down another rapid by line. Pulled out, ran across a riffle. Struck in, going over, landed at the head of another rapid on the right. Boats let down by line, then got in and pulled for 3/4 mile in smooth water. Landed on the left bank at the head of a rapid. Let down by line. Camped on the left at the foot of the rapid for dinner. Water fall some six feet in 1/4 mile. After dinner pulled out for 1/4 mile, landed at the right bank. Prof. climbed out while we waited for him. It rained, felt chilly. The first rain we have had for three weeks. Rain continued only half an hour. Beaman took some pictures of a vertical cliff. Waited for him two hours, then pulled out and ran three rapids, two bad ones. Landed at the head of a fall, let down by line [Figure 5]. Fall of water six feet. Pulled out for half mile, landed on the left bank to examine another rapid. Pulled over to the right and camped for the night, under some cotton wood trees. 6-1/4 miles.

[Aug.] 22. Pulled out this morning. Commenced with running a rapid half mile long. Made it in just two minutes. It was in the shape of a third of a circle. Landed on the left at the head of a fall. Let down by line, four men guiding the boats through the rocks. Pulled out through another rapid, landed at the foot to let down over another. Pulled out for ¾ mile. Landed at the right bank at the head of a rapid. Ran through it partly, then landed and let down the rest of it by line. Pulled out for a mile and a quarter in smooth water. Landed on the left at the head of a fall. Let down by line, four men guiding boats. Passed rocks, camped for dinner at the foot of a fall on the left. Pulled out at 3. Ran a huge rapid. In going over the Dean struck twice. Jumped out and pushed her over. No damage. The water, in rushing through a narrow shoot, caused the waves to boil up some six feet. Ran another about three quarters mile farther down, then landed on the right bank. The country all of a sudden opened out on the right. Saw an Indian pony turned out to die. No Indians being in sight, and the horse being lame, causes us to conclude that he had been left. At first we thought that the Major might have come down here, but the manure seemed to be very old. Pulled out, ran a rapid, landed at the root to examine another, which we concluded to run. Pulled through and landed on the right bank at the head of a fall. Camped for the night.

August 23, 1871

[Aug.] 23. Prof. and Jones climbed out and returned at dinner. Took in a feed of sowbelly in our stomachs and loaded the boats with our baggage. Let boats down by line over the falls. Called them Sharp Mountain Falls. Beaman photographed an over hanging cliff of stone 1200 feet in height. Pulled out for a mile in smooth but swift water. Ran a rapid half a mile long, landed at the head of another. River split up in two channels. The Nell took the left but ran aground just ahead of the rapids. Prof. told Jones to take the right channel, which we did. An old tree had fallen off from the bank and lodged. The current passed through this tree, and not being aware of the great suction, was carried into the tree, struck on Fred['s] oar lock, which it snapped like a pipestem, and there hung under the tree. I removed my oar and lock then pushed clear and we went at the rate of 30 miles an hour for a quarter. Landed on the left bank ¾ mile farther down and camped for the night. While cutting some willows for my bed I discovered a huge rattle snake coiled up on the roots of willows. I called to some of the boys to bring a pistol, when Steward brought the Prof's, and shot him through the side of his belly, but a second ball passed through his head. I threw him on the beach. Counted nine rattles and a button, which made him nearly ten years old.

August 24, 1871

[Aug.] 24. Pulled out this morning in smooth water for ¾ mile. Ran a rapid and then ran in quiet water for nearly two miles. Landed at the head of a fall on the left but pulled to the right and let down by line. Got an awful rap on my shin. Lost a handkerchief and bandage which was around my leg. Fall of water 10 feet. A nice little creek coming in just above the falls on the right started by some springs some 10 miles above its mouth. Pulled out, ran the rapids then had a quiet river for nearly 3 miles. Landed at the head of a rapid on the left. Shot at some otter but failed to get any, being the first we had seen on the trip. Let down by line. Water fall about 7 feet. Pulled out into another rapid but failed to make a clean run. Struck as we went in, jumped out, pushed over and then pulled through for ¼ mile. Ran another a little below, and still another about 100 yards below. Made them all OK. Landed on the right for dinner, pulled out at 2:15 and ran a rapid immediately after starting. Went whooping for a mile, then another rapid. Ran it then for a mile and a half in quiet water. Landed on the right at the head of a fall. Let down by line for little more than one quarter, pulled out for ½ mile, then ran a rapid. Pulled out and landed on the left at the head of a rapid. We have seen coal all day. Concluded to be out of the Canon of Desolation, and are now in the Coal (Lignight) Canon.[20] The features of this are the same as in Desolation. It is

[20] Also called Gray Canyon.

God's country, for man don't want it. The walls in this canon are narrowing up again. The walls are perpendicular but not high. Ran the rapid which proved to be an old hummer. Made it all right. Ran another, then had shallow water for ¾ mile, then ran another rapid. Landed on the right to examine one which we ran. Struck as we went over. Landed on the right. Camped for the night.

August 25, 1871

[Aug.] 25. Prof., Jones, and Capt. climbed out, returning at 11:30. Reported that we were running out of canon. Had dinner and pulled out. Ran two rapids immediately after leaving camp. Landed on the right at the head of a fall. Let down over two falls. First fell about six feet and the second five. Ran along for a mile, then came to another rapid. Ran it and four others all OK. Came to the mouth of the Little White [Price River], about 60 ft. wide, but at this season of the year contains no water. We placed our beds on a shelf of the river bank. The Valley of the White is rich in scenery. It is called Castle Valley. The mountains on each side are so much eroded that they form all sorts of castles and fortresses, hence the name. Immediately on the bank of the river grow box elder, cotton wood, willows and grass.

August 26, 1871

[Aug.] 26. Pulled out this morning. started with a rapid half a mile long. Ran it all right. We shot over it like an express train. Pulled a half mile in smooth water for a mile and a half. Landed on the right to examine a rapid. Ran it little farther down. We found two more close together, two old hummers. Ran them like a demon. The Nell and Canonita shipped half full, but the old Dean, happy-go-lucky, went over without any. Little farther down we ran another, landed on the right to examine a rapid. Proved to be an old buster, but ran her all right. Ran along for two miles, then ran another, Little below, we landed on the left to examine another. From this point we can see the end of Coal Canon. Ran it all right. Just below the rapids Capt. shot a beaver. Had hold of him twice, but each time slipped out of his hand, when he finally sunk. Little farther down had a rapid which proved hard work getting the boats over. The river at this point covers nearly one quarter mile in width. Worked nearly an hour. Ran little farther, then camped on left bank just inside the canon. Have to wait for the Major till the 3rd of Sept. After dinner hauled up our boats to let them dry for caulking.

[Aug.] 27 Sunday. Dried rations and fixed my clothes. Fred and myself went down to plant our flag at the foot of an adjacent island as a signal for the Major.

August 27, 1871

[Aug.] 28. Caulked and pitched boats. Caught three large fish, one weighing 15 pounds. Had him for supper.

August 28, 1871

[Aug.] 29. Commenced fixing my moccasins. At dinner Beaman, Fred, Clem, Hatten and myself were discussing the Major's probable return, when all of a sudden three shots were heard—his signal. Everybody was up in a moment, answered the signal, and in a jiffy Prof., Bish and myself were across the river with a boat. A few minutes after we saw the Major and Mr. Fred Ham[b]lin on the brow of the hill. Soon pulled them over to our camp. Major gave orders to pack up and start down some five miles where he had left the pack train. Soon were skimming over the water, while Jones went round by land with the horses. Ran three rapids. The Nell, in going over, got foul of a ledge of rocks but sustained no damage. Arrived at Gunnison's Crossing, so named after Captain Gunnison, who in '53 crossed the river at this point to explore the country around here, and farther west

August 29, 1871

in Utah he was killed by Gosi Ute Indians at Severe [Sevier] Lake, for a long time called Gunnison's Lake. He had strayed a little ways from his party when they killed him. Found two letters for me from my brother in Frisco.

August 30, 1871

[Aug.] 30. Spent the day writing and tinkering.

August 31, 1871

[Aug.] 31. Major, Beaman, Clem and myself went out with the instruments to take a picture of Gunnison's Butte. Returned at dinner and immediately went out again to take more views. Returned at 5 PM.

September 1, 1871

[Sept.] 1. Mr. Ham[b]lin and his nephew made preparations for returning to the settlements. At noon they left, taking with them our mail, fossils, and other loose baggage. Soon after we dropped down stream half mile, being camped on the right. Moved over to the left in order to be able to get back to our last camp in the canon where the Nell had left a saw. Beaman and Fred went after it. Returned at dark. Camped for the night.

September 2, 1871

[Sept.] 2. Pulled out at 8 AM. Ran along through Gunnison's Valley. Came to many shoals. Jumped out very often to put boat over, but no rapids. Camped on the right bank for dinner. Immediately after pulled out and stopped for barometrical observations. At the Spanish Crossing at 2 PM. Someone had crossed here lately, as fresh horse tracks could be seen on the gravel, also a barefooted boy's print in the sand. Pulled out soon after observation. This trail formerly was used quite extensively by the Spaniards in going from Santa Fe to Los Angeles, Calif. Ran down some four miles and camped on the left side in a cotton wood grove.

[Sept.] 3. In camp. Major and Jones went across and went to some bluffs about 9 miles from here. Returned at night. Major brought a lot of fossils. Steward and Bish. went out some four miles. They found a cave of some two hundred feet in length. Made myself a pair of moccasins.

[Sept.] 4. Pulled out at 9 AM. Still running through what the Major calls Gunnison's Valley, but now and then would pass through a line of bluffs, highly colored. Made ten miles then camped for dinner on the right bank under a sandstone bluff. Country is barren and lifeless. Pulled out again at 2 PM. Saw a beautiful little butte in the shape of a cathedral. Beaman took a photograph of it. Major called it "Dellenbaugh's Butte" after Fred. Ran down some 5 miles when we came to the mouth of the San Rafael, a stream of about the size of the White. Starts at the Wasatch Mountains. Camped on the right hand.

[Sept.] 5. Major and Jones started up the San Rafael to be gone two days. Boys found a lot of arrow heads. This part of the country is not much traveled at this season of the year as the game is all upon the mountains, but in the winter the valley is a good place for game.

[Sept.] 6. Major and Jones returned at 10 AM. Bish and Clem started out at 7 AM for a butte some ten miles from here on the left bank. The peaks which the Major and Jones visited are called "Jones Peaks." This part of the country is not laid down on the map and is left blank, but on our return will be filled up. Helped Jones to measure the distance from mouth of San Rafael to Jones Peaks. Found them to be ten miles. At this place the Indians make their arrow heads as we found their tools, stones, and a lot of chips. Bish and Clem did not return. Kept a fire burning all night as a signal for them, thinking that they had lost their trail.

September 7, 1871

[Sept.] 7. Major and myself started out in search of Bish. and Clem. Found them on their way in some five miles from camp. Struck for the river where Prof. had come down with the boat. It began to rain fearfully when we reached the river, and a little while after hundreds of streams came shooting from the top of the bluffs. One in particular came shooting over a vertical wall 200 feet high. The spout was some eight feet in diameter. It looked like a river of red mud rolling down, having come from flats of red sandstone dust. It was a grand sight. We took shelter under an overhanging wall. Had dinner and then pulled out. Still raining. All the way down streams came shooting down, some bright as cristal while others were of a reddish or umber collor. Camped on the left bank in an oak grove. First oak seen on the trip. At noon we ran into what the Major called Labirinth Canon. Walls are increasing in height as we go down. A very peculiar feature is noticeable: the wall on one side is vertical while on the other it is rounded and sloping back. It changes at every bend of the river. In the bend the walls are vertical while on the other they slope and continue so until the river makes a turn. Stopped for half an hour, for Fred to do his sketch, being unable to do it in the rain.

September 8, 1871

[Sept.] 8. Beaman took a photograph of a tripled alcove [Trin Alcove], and after climbed up on top of a bluff from which he took several pictures. In the afternoon crossed over and took several more.

September 9, 1871

[Sept.] 9. Pulled out about 9 AM leaving the Nell and the Canonita behind. Landed about two miles from camp for the Major to geologise. Pulled out after stopping half an hour. Ran several times aground. About noon it began to rain. Landed on the left. Pitched our paulin, built a fire, and waited the coming of the other boats, who soon made their appearance. Had dinner and pulled out. Walls increasing in height as we go down. About 4 o'clock the Dean got into a channel which proved hard work for us. Ran aground, signalled the other boats to take the left hand channel. All the Dean's crew got out to push her over. Dragged her for some twenty-five feet but being unable to get her any farther we had to call on all hands. At the end of the bar we had only about three inches of water, so we had to lift her over some twenty yards more. Ran until 6 PM. Camped on a sand bank in bow knot. Made 16 miles. The river at this point bent back and forward forming a bow knot. Major called it Bow Knot Bend.

[Sept.] 10. Beaman climbed up on top of a ridge about 300 feet high and a thousand wide, and four miles long, round which the river doubles, forming the last loop of the bow knot. Pulled out at 10:30 leaving Beaman on the ridge photographing. To meet us on the other side. Got on the other side a little before 12 Noon. Camped on the left for dinner. Concluded to stay for Beaman to fix his chemicals. Major and myself climbed out on the ridge while Prof. Jones, and Steward climbed out on the left.

September 10, 1871

[Sept.] 11. Pulled out at 8 AM. Ran down some six miles and camped on the right bank for dinner. The walls of the canon from 800 to 1000 feet in height. Very much cut up, forming numerous alcoves, pinnacles, columns, peaks, towers, castles and cliffs. Water placid and very shallow. Pulled out at 3 PM. While preparing dinner the boys dried out their blankets which had got wet during the last night's dew. Stopped on the right for Beaman to take a picture of a latteral canon. While

September 11, 1871

some of us climbed up, the Nell pulled out to a place where the canon breaks down some five miles from here. As soon as Beaman had taken his view we started after and overtook them just as they landed to camp, being six o'clock and getting dark. Camped on the left.

September 12, 1871

[Sept.] 12. Pulled out at seven o'clock. Ran down some seven miles. The walls of the canon are broken down very much and I think can hardly be called a canon, being in some places only about 200 feet, while on the other points it is level with the river excepting a bank. Saw a curiously eroded mountain or rather a cliff. It was in the shape of a Greek Cross [Butte of the Cross]. Beaman photographed it on the right bank. Major, Prof., Steward, and Bishop climbed out on the left while we dropped down some 400 yards to prepare dinner [near Barrie Creek], leaving Beaman to come round the point by land. The whole party united at one PM. Had dinner. The Major, Beaman, Jones, Fred and Clem took the Dean downstream about a mile, while we put up our cooking utensils. Soon followed Andy and myself taking down the Canonita. Major came down soon. Started down with the Dean. Prof. came down with us some three miles and went into camp on the right bank. At this place the river has cut through a wall of sandstone round which the river formerly flowed, making a cut off of some four miles. The bend [Bonita Bend] on each side was to the side of the wall. The old riverbed is distinctly visible. Left our axe at dinner station. Major concluded to call this the end of Labyrinth Canon. The break in the walls extends over some nine or ten miles.

September 13, 1871

[Sept.] 13. Major and Prof. climbed out, while Beaman took some views. Pulled out at 11:45. Ran four miles and a half. Camped for dinner on the right under a shelving of rock. While eating dinner heard the bark of a wolf a little ways down in the bend of the river. Some of the boys mistook it for the bark of a dog, thinking it might be from some Indian camp. This part of the canon is very wide, forming quite a little valley on each side. Walls are overhanging—that is, the homogenus sandstone projecting over the red sandstone which came up at last night's camp. After dinner pulled out for a mile and a half, when the Major and Prof. climbed out. Beaman took a picture of a valley of rocks. I never saw anything to equal it. It was nothing but peaks, pinnacles, towers, spires and chimneys. Pulled out again at 4:30. Ran down about seven miles, then camped on a high bank. The canon since four o'clock is narrowing up—perpendicular and very ragged. Height 800 to 1000 feet. The country back is cut up, forming numerous terraces, gulches, alcoves and winding stairways.

[Sept.] 14. Major and Steward hunted for fossils. Beaman took a picture of the canon, from a bend a little below camp. Major, Prof. and myself pulled up stream half of a mile and obtained a lot of fossils. Returned at noon. After dinner pulled out but ran but a little ways when the Major got into a fossil fever. Stopped for 45 minutes. I found a large fossil tooth which I gave to the Major. Pulled out till four PM. Landed on the right. Boys found a lot of Moqui ruins, perhaps, some 200 years old. This ancient race has been driven from country to country on account of their industry—cultivating the land and raise enough to subsist on. Make pottery, blankets and everything that they require. The other tribes don't like them because they work.

[Sept.] 15. The Nell pulled out for the junction of the Grand and Green for Prof. to get fornoon's observations. Major, Jones, Hatten and myself climbed out

September 14 ,1871

September 15, 1871

73

this morning. While going up a gulch Hatten found an Earthen pot full of split willows and reeds, the work of Moqui Indians. Had a hard time getting out onto a ledge of rock on which we had to get out in order to reach the top. At that place the Indians who resided here used to climb out, for two cedar trees were set against the wall looking old and decayed. By their assistance, I should judge when they were green, could reach the ledge. By the assistance of some rocks which were also piled by the Indians, we commenced our ascent. Jones, being the tallest, got on my shoulder, and so reached the shelf. Next came the Major, whom Jones pulled up by a rope brought for the occasion, then Hatten. Myself, having moccasins on, I shimmied up like a lizard. After reaching the top walked out two miles to a little butte, where we took bearings, and returned at noon. At the shelf we lowered ourselves down with the rope. I was the last. Lowered myself as far as I could when Jones pressed my feet against the wall and slid down. After dinner pulled out, reached the junction at 4 PM. The Grand is a stream about the size of the Green and at present has more water. United they form the Colorado. At the junction a Mr. ———— of Colorado is supposed to have made a treaty with the Indians for the possession of this valley, with a site for a city at the junction. The government furnished Mr. ———— with necessary funds to carry out the treaty, which was supposed to have been signed by the great chief, Black Bear, and a few other prominent chiefs of tribes, under a cottonwood tree. But where the valley, the site for a city, or even the cottonwood trees, I can't see. The walls at each angle come close to the rivers, leaving a narrow bank of drift sand on which a few willow trees seem to flourish. Height of walls 1300 feet, nearly vertical. At each angle the strata is turned up making the dip nearly 25 degrees inward. Major concluded to call the last canon from the break in Labyrinth Still Water Canon.

[Sept.] 16. Major, Jones, Beaman, Fred, Clem and myself pulled up the Green some ¾ mile and climbed out. Beaman took a lot of pictures. Fred Sketched. Clem and myself gathered a lot of pitch which found to be plentiful. From this point we could see the Sierra La Salle stretching northward. A grand view is presented to the eye from this point, of craggs, buttes, spires, and pinnacles in various shapes and forms and color. Major and Jones returned at 3 PM. Reported as having seen parks formed or enclosed by pinnacles. Returned to camp, leaving Beaman's apparatus on top. Got two gallons of pitch.

September 16, 1871

[Sept.] 17. All hands excepting Bish and Jones started up for to see the parks. After an hour's climb we reached the top then walked two miles to the parks. Such a sight I shall never forget. I counted five parks enclosed by pinnacles formed by erosion. They looked to me like monuments in a cemetery. Everything looked sombre and deathlike. Nothing disturbed the scene except the sighing of the wind or the falling of a chip from the rock. The water collects in a large basin in the center of each park and from hence into a gulch into the river.

September 17, 1871

[Sept.] 18. In camp. Major and Jones climbed out on the left angle of Colorado and Grand. Returned late at night. A breach of some 50 feet wide was visible on the right of the junction, but the current cut it away in three days.

September 18, 1871

[Sept.] 19. Pulled out this morning at 10. Divided all the rations before starting. Ran down some four and a half miles when we came to a huge old rapid. Landed on the left and let down by line, each crew working their own boats, Fred and myself wading by the side of the Dean while Jones held onto the rope from the junction. This canon is called "Cataract", and well it deserves its name. The water, while going over a fall or rapid, it goes down like steps of stairs. Let down one more

September 19, 1871

before dinner. Immediately after dinner let down another half mile farther—another old buster—let down for half mile—then ran a half to the head of a fall, let down, got in, pulled for ¾ mile when we ran one picking our way through rocks. Also ran two small ones. Came to a fall, let down and picked through the rocks. Got in and pulled through two almost joining. Went into camp on the left.

September 20, 1871

[Sept.] 20. Major and Fred climbed out on the right. I took them across at the head of the fall. Bishop and Steward went out on the left for topography and geology. Prof., Jones, Clem and myself let the Nell down by line on the right side, Prof. and Jones at the rope, Clem and myself guiding her through the rocks. About half mile down we came to some rocks which projected out into the river from shore, immediately below an eddy. Being afraid that she would run up under the fall and fill, so we attached the other end of our rope to the stern for Prof. to pull her in short, while Jones, Clem and myself held onto the bow. Down she went. Prof. pulled her stern in a little too much, which swung her bow out, and in a jiffy the current caught her bow, and no ten men could have held her, the rope being too short, or else we would have held her. After turning we had to let go or be dragged over the rapid. We choosed the former. Soon the rope tightened in Prof's hold. He being unable to resist the pressure, let go too, and the Nell went fluking over the fall, while we kept up the race with her on shore over the rocks. She went over all right. Immediately below two more falls were visible. Fearing that she would go over these, we ran for dear life—visions of short rations could be plainly seen. Prof. started in but backed out. I plunged in and swam to her almost breathless. I gained her deck twenty feet from the head of the lower fall. I reached an oar to Clem whom I pulled in. A counter current took her far enough for me to reach an oar to Prof. who pulled us in just in time for to save us going over the falls below. Went back and let the other two down on the left all right. Camped on the right on a sand beach. Major and Fred returned about 5 PM.

[Sept.] 21. Pulled out at 8. Good river for ¾ mile then came to a rapid. Ran it and five others, then let down three old hummers, four men guiding her through the rocks. Ran two more, then camped on the left at the head of a fall for dinner. About 1:30 we let down by line. Walls of canon from 1000 to 1700 feet high, very ragged, composed of limestone, and are still increasing in height. About a mile farther down we had another let down. Wading by the side of the boat, sometimes wading while at others hang[ing] on for dear life. Took us nearly two hours. Immediately below another. Major, seeing how wet, cold and tired we were, concluded to run it at a great risk. Pulled out into it, but no sooner in than we struck a rock, nearly keeled over, but fortunately slid off. Pulled for half mile through foaming cataracts. Shipped our standing room nearly half full of water. Camped on the right on a sand beach.

September 21, 1871

[Sept.] 22. Commenced by running a small rapid then let down two. At the first, had to get the boats out of water and slide them over rocks. Broke our keel plating. Worked all forenoon to make two portages. How Mr. White[21] came

September 22, 1871

---

[21] James White, a trapper, appeared at Callville (established in 1864 by Anson Call at a point on the Colorado River just above the present site of Hoover Dam) below the Grand Canyon on September 8, 1867, claiming to have escaped from Indians, built a raft, and floated down through the Grand Canyon for eleven days. He was starving, badly sunburned, and demented. His story was printed in January 1869, in the *Rocky Mountain News*. There is some evidence that Powell sought him out and questioned him about the trip in early 1869 (Wallace Stegner, *Beyond the Hundredth Meridian: John Wesley Powell and the Second Opening of the West* [Boston, 1954], pp. 33, 376). Robert Brewster Stanton in his *Colorado River Controversies* (New York, 1932), pp. 70–93, investigated the story at length and concluded that White had perhaps floated from the Grand Wash Cliffs, below Grand Canyon, to Callville.

through this canon on a raft in one day is more than I can, but such is laid down in history, but I don't beliefe it. A raft could not live—it would go to pieces while going over the first cataract. Dinner-camp on the left, under a hackberry tree. Walls opposite vertical and 200 ft. high. Let down another cataract. Took us until 4 PM. Went into camp on the right—sandy beach. Repaired the Nell and Dean.

September 23, 1871

[Sept.] 23. Let down over an old snorter. Ran a half mile, let down another. Got in and pulled through one all right. Stopped at the head of a rapid. Took out the cooking kit and let down while Andy prepared dinner. I took the Major down in the Dean to head of a fall, tied up and returned to camp where we had to wait for the boys, having left them below the first fall, for Beaman to take pictures. Soon the Canonita came in. Had dinner on boat. Dropped down to the head of a fall. Major concluded to run it, which we did in fine style. Beaman took an instantaneous view of us as we went over. Fall of water, 12 ft. in 50 yards. Pulled to the head of the next fall, where we waited for the other boats. Had a mile of swift water. Walls very craggy and vertical on top. The other two boats soon came. Crossed over to the left and let down by line for half mile. Hard work getting boats over. Took us the rest of the afternoon. Fall 70 ft. in half mile. Camped on the left at the foot of fall.

September 24, 1871

[Sept.] 24. Pulled out this morning into a rapid, made it all OK, but to be stopped by a fall. Let down by line, stationed all hands but two along the fall below, with the rope to pull her in when she came over. Two men shoved her off. It put one in mind of a bucking mule to see the boats jump through the waters. Got in and pulled out for a mile and a half through rapids. Done some wild riding through the rocks but got through all OK. Stopped at the head of a long rapid, let down for half mile. In going round a big rock the stern of the Dean caught on a sunken rock and nearly capsized and filling her half full of water. With the assistance of two men, slid her off, got in and pulled through a rapid half mile long, made it all OK. Little farther down ran another, then came to one where we let down. Being full of rocks, got in and pulled through another. Landed at the head of a cataract. Let down by line just below. Went into camp on the left, hauling our boats up to dry. Had dinner, then fixed around my clothes and shoes. Major, Jones and Beaman had a walk up a lateral canyon [Gypsum Canyon]. Returned late at night.

[Sept.] 25. Major and Fred climbed out for observations, while Beaman, Jones and Clem. went up the canon for pictures. Bishop working at his map. Prof. and Steward pitched the Nell and then went up the side canon. I pitched the Dean and helped Andy fix the Canonita. All but the Major and Fred returned at 5 PM.

September 25, 1871

[Sept.] 26. Pulled out this morning to the head of a cataract little below camp. Let down by line. Got in and pulled through into rapids, ran a little ways, then landed on the right to examine another. Walls craggy and slanting back, towering up some 3000 feet. Pulled out into the rapid, and went fluking for half mile. The Nell while coming over, filled but managed to reach the shore in safety. The Canonita struck and half filled, but sustained no damage. Little below, about quarter of a mile farther down pulled through a rapid about one-third of the way, then let down another third, then pulled through the rest to the head of a fall. Stopped for examination. Ran it all OK. Shipped some water. Fall of water 7 feet. Pulled out and ran two small rapids. The canon is very narrow, walls perpendicular, from two to three thousand feet. Broken up in some places, forming towers and

September 26, 1871

pinnacles, some of them nearly 1500 feet. Limestone at the bottom and sand on top. The former causes so many rapids, being the hardest. Landed on the left for dinner. Immediately after dinner pulled out into a rapid. Made it all right. Waited on the left for the Canonita, then ran another small rapid. Landed on the right and went into camp. Beaman took a view of the canon.

September 27, 1871

[Sept.] 27. Went up a gulch [Dark Canon] near camp for Beaman to take pictures. Returned at 10. Let down over the rapid at camp. Immediately after ran down for a mile and had dinner. Pulled out after dinner for a half mile, then came to a cataract. Let down by line, got in and pulled through a rapid just below the cataract. Ran another and then had a beautiful river for nearly four miles, then came to a cataract. Walls dropping a little in height. Ran it all right. Had three miles more of good river with three rapids in it at equal intervals, then came to a cataract. Let down by line and went into camp just below. Made nine miles.

September 28, 1871

[Sept.] 28. Pulled out this morning leaving the Canonita for Beaman to take pictures. Just below camp ran two rapids. Below the second we left the Nell to signal Beaman how to run. Pulled for a mile in nice river, then came to a huge cataract. Waited for the other boats who soon made their appearance. Major and Prof. climbed out but failed to get out. Returned about 4:30. Let down by line. After supper pulled out, ran a rapid just below camp, about a quarter of a mile farther down. Landed and let down through the head of another. Then had to sink or swim, for no footholds be had on either side. The current [illegible] is very strong toward the left wall and came near smashing us to atoms, but saved ourselves by a few feet. The other boats did not get quite so close, being aware of the strength of the current. Pulled out and ran two more. Darkness came on and we went into camp at the head of a rapid on the right bank.

[Sept.] 29. Had a terrible night of it. Wind blowing the drift sand like snow, and everybody had red eyes this morning. Rained hard this morning. Delayed us for an hour, then pulled out into the rapids. Made it all OK. Raining all forenoon off and on. Ran down about five miles to Mille Crag Bend. Major and Prof. climbed out. This is the end of Cataract Canon, and thank God, [for] it has been hard work ever since we left the junction. The walls break down and are only about 1800 feet high, all cut up into pinnacles and towers, crags innumerable, hence the name. Walls are sandstone. Camped for tonight on the left. Ran the limestone under and now [f]ace the Homo G[eneous?] at the top and bottom.

[Sept.] 30. Pulled out at sunrise. Ran a rapid just below camp, another a little ways from the first. Ran about a mile. Ran two more, then ran for six miles in a beautiful straight canon, no rapids. Walls on both sides come down to the river and are only about an eighth of a mile wide. Called it Narrow Canon. Came to the mouth of the "Dirty Devil" River, the end of Narrow. The Dirty Devil is a stream of about fifty feet wide. Water very muddy. It starts from a range of volcanoes [Henry Mountains] about forty miles from here. Not laid down on the map. Major, Prof and Jones climbed out. Returned at night. The walls break down to nothing, running the sandstone under.

Oct. 1. Cached the Canonita until next spring, having no material for photographing and are short of provisions. Major intends to explore this part of the country, bring in material and ride down in her on the Colorado to the Par Weep [Paria River]. Hauled her up into a cave about 20 feet up from the river and covered her up with sand. Major and Jones climbed out to find a trail leading in. The Nell dropped down about a mile and a half for them to climb out from there. About 4 PM Jones and Major returned and we dropped down to where the Nell had landed. Went into camp.

October 2, 1871

[Oct.] 2. Pulled out early this morning. Ran two rapids just below camp. Struck in the second, had to get out and push boat off. Water very cold. Wet clothes all day. Running up the red sandstone. Major calls them the "Orange cliffs." Landed on the left to examine some old Moqui ruins [Figure 6]. Found one house about 20 ft. long, and the walls, what yet remained, were 14 feet high.[22] Found a lot of hieroglyphics on a smooth sandstone. Could make out a good many things out of them but none perfect or recognizable. Had dinner and pulled out for 11 miles. Camped on the left where we found another old ruin.[23] Had seven rapids, ran them all OK. Called this Mound Canon.[24] The top of the plateau is curiously eroded. Look like huge mounds closely joined, like a grave yard.

[22] This ruin stood on a prominent point overlooking the mouth of White Canyon. It is described in Ted Weller, "San Juan Triangle Survey," in *The Glen Canyon Archeological Survey*, University of Utah Anthropological Papers no. 39, pt. 2 (Glen Canyon Series no. 6) (Salt Lake City, 1959), pp. 543–669.

[23] This ruin was excavated in 1958–59 by the University of Utah (William D. Lipe, *1958 Excavations, Glen Canyon Area*, University of Utah Anthropological Papers no. 44 [Glen Canyon Series no. 11] [Salt Lake City, 1960]). The University named it Loper Ruin for Bert Loper who built a cabin near the site in 1908 at the mouth of Red Canyon (C. G. Crampton, *Standing Up Country: The Canyon Lands of Utah and Arizona* [New York and Salt Lake City, 1964], pp. 164–65).

[24] Powell originally called the canyon from the Dirty Devil to the San Juan rivers "Mound Canyon," and the stretch from the San Juan to Lee's Ferry "Monument Canyon." He later combined them into "Glen Canyon."

[Oct.] 3. Pulled out at 8 AM. Ran till noon. Made 14-3/8 miles. Had a fine run, only a few rapids. Pulled out at 2 PM, and ran 13-3/8 miles. Camped on the left. Canon very near a mile wide. Walls running up and down from 100 to 700 feet in height. Looking from the top it is a vast stretch of sandstone eroded, forming huge mounds closely packed together, like a grave yard. No vegetation can be seen, nothing but the bare rock. What the Indians would call [Unkar] to weep,[25] or Red stone Land. Huge water pockets can be found all over, varying from two to thirty feet in depth. In one we found a cotton wood tree. It was a well half full of sand or nearly so, judging from the appearance.

[Oct.] 4. Pulled out about 8 AM. Ran for two miles on a beautiful river, then came to shoals, or rather we came into a bed of hetrogeneous sandstone. The bed was as smooth as a barn floor. Had to slide the boat for nearly two miles, then changed directions and ran it up. In coming over a small rapid the Nell stove a hole in her centre cabin. About noon hauled her out and repaired her and had dinner. Pulled out 2:25 PM. Ran nearly 20 miles. Camped on the right under a ledge of rock.

[Oct.] 5. Pulled out this morning 7:20. Ran five miles and a half. Landed on the right at the mouth of a small creek. Tasted strongly of alkali. Pulled out again after delaying half an hour. Passed the San Juan about noon. Stopped two

[25] Hillers probably learned this term from Powell. The Ute and Southern Paiute vocabulary lists collected by Powell between 1868 and 1873 contain several entries listing "Unkartoweap" as "Red stone land." (Don D. Fowler and Catherine S. Fowler, eds., "The Anthropology of the Numa: John Wesley Powell's Manuscripts on the Numic Peoples of Western North America, 1868–1880," *Smithsonian Contributions to Anthropology*, vol. 14 [1971]).

October 3, 1871

October 4, 1871

October 5, 1871

miles below the mouth. The San Juan starts from the San Juan Mountains on the western slope of the Rockies. At dinner went up into a gulch, about a quarter of a mile, when our progress was stopped by a large temple or space containing a small pool of water [Music Temple]. The water had undermined a solid rock, making a regular amphitheatre of about 500 feet long, 300 wide and 200 feet high. The walls rounded up, almost closing at the top, leaving only a space of about 20 feet for daylight. Before entering this temple a beautiful little cotton wood ornamented its portals. At the junction of the San Juan and Colorado Mound Canon terminates.[26] Pulled out after dinner. Came in view of a volcano [Navajo Mountain] on the left. Major called it Mount Seneca Howland in memory of his old topographer. It is about 4000 feet high and about 14-15 miles long. Ran four bad rapids scattered at intervals. Canon very wide. Off from the river numerous huge piles of rocks can be seen, towering up to 1500 feet. Look like monuments. The Major calls this Monument Canon. All along the walls the rock has been eroded by the drippings of springs, making it sloping, and on this slope grow small oak and flowers, forming little glens. They are very numerous. Camped on the left. 25 miles.

October 6, 1871

[Oct.] 6. Pulled out this morning. Had a fine river. Found some newly burnt brushes, the tracks of a horse, and white men. Concluded that it must be the Ham[b]lins looking for us, having come down from the Crossing of the Fathers, or some disappointed prospector. About half an hour before we camped for dinner we found a fox on a few rocks in the center of the river in a rapid. Boys fired at him when he took to the water and was carried down the current. Saw him

[26] See diary note 24.

84

no more. Camped for dinner on the right on a bench of rock. Pulled out after dinner for the Crossing of the Fathers at which point we arrived about 4 PM. Heard a gun, watched the bank for a moment, saw a little white streamer attached to a pole, and presently saw my old friend Capt. Dodds, of Uintah acquaintance. Landed and found rations, but Mr. Hamblin had gone to Ft. Defiance, Arizona, on a Mormon mission, to obtain indemnity from the Government—the Navajos having been down to Kanab and forced the people to give up their stock and everything else they wanted. They submitted rather than have war with them. The Navahoes receive an annuity from the Government. Hamblin hopes to stop some of it. Found also two prospectors having come down with Dodds from Kanab.[27] They were anxious to reach some canon where the water ran swift. They said that they had come all along up the river from the Virgin. Found gold everywhere, averaging from 3 - 5 dollars a day.

[Oct.] 7. In camp. Fixing up for a trip to Salt Lake with the Major.

[Oct.] 8. Still [in] camp. About a half mile round the bend is the old Ute Crossing, or more largely known as the "Crossing of the Fathers." Ascalante [Escalante] crossed with 100 pri[e]sts to be distributed among the different tribes. Nothing was ever heard from them afterward. It was supposed that the Indians killed them.[28]

October 7, 1871

October 8, 1871

[27] Jones ("Journal," p. 98) lists the prospectors as George Riley and John Bonnemort.

[28] Hillers may have heard this "tall tale" from Riley or Bonnemort. The Domínguez-Escalante party of 1776 contained only ten persons, Domínguez and Escalante being the only priests. The purpose of the party was exploratory and not to distribute priests among the Indians. The party returned safely to Santa Fé in late 1776. See Herbert E. Bolton, "Pageant in the Wilderness: The Story of the Escalante Expedition to the Interior Basin, 1776, including the Diary and Itinerary of Father Escalante," *Utah Historical Quarterly* 18 (1950).

October 9, 1871	[Oct.] 9. Still in camp. Boys writing letters.
October 10, 1871	[Oct.] 10.[29] Started out this morning with pack train. Capt. Dodds and the two miners started back with the Major and myself. Followed the trail for about 12 miles. Went into camp near a gulch. Found water for our animals—full of alkali and salt.
October 11, 1871	[Oct.] 11. Started out this morning and went about 12 miles to a spring of alkali and salt. One of the miners and myself started up the mountain for some game but found none.
October 12, 1871	[Oct.] 12. Started out early. Reached the Pa Weep [Paria River] about 5 PM. having come about 25 miles. The river is narrow and very swift Good tasting water. Starts from Wasatch Mountains. Wind blew cold.
October 13, 1871	[Oct.] 13. Major and Capt. started for Kanab, leaving us to bring the train. Started with the pack about 8 AM. Shot four ducks. Made about 16 miles. Left the trail and ran up a gulch for two miles to a spring and camped.
October 14, 1871	[Oct.] 14. Duck for breakfast, and pumpkins which we had brought from the farm. At the Pa weep where the Mormons try to make a settlement [Paria] having built a few homes, walled in. Made Kanab about 6:30, having come 30 miles.

[29] Hillers's diary for October 10 through November 30, 1871, is the only extant account of Powell's trip to Salt Lake City to bring Mrs. Powell, his infant daughter, and Ellen Thompson to Kanab for the winter.

Meanwhile, Thompson and the rest of the party continued, arriving at the mouth of the Paria River on October 28, 1871. On November 6 they started overland to Kanab, arriving there on November 12, 1871 (Thompson, "Diary," pp. 56–61).

[Oct.] 15. Kanab is a place of about 25 settlers, on the Kanab River. Kanab means "willow" in Indian. It sinks in the sand about six miles down the valley. Had a long ride this morning after our stock, returned at noon. In the afternoon the two miners dissolved partnership. Riley was hired by the Major to take rations down to the mouth of the Pa weep. Beaunamat [Bonnemort] intends to go to Beaver.

[Oct.] 16. Fixed wagon covers and got ready for starting tomorrow. Major bought Bonemote's outfit and he goes to Salt Lake with us.

[Oct.] 17. Hitched up our two mules and sailed out Took road by way of Scum pa [Skutumpa] the Indian for it. It means Rabbit bush water. Went thirty miles to J. D. Lee's farm, who entertained us hugely.[30]

[Oct.] 18. Started this morning for the head of Severe [Sevier River] which place we reached at night. 30 miles. Camped by a large spring by which this branch of the Severe is started.

[Oct.] 19. Had a terrible storm last night. Our blankets wet through. Found my left side in water. Top blanket froze stiff as a poker. Started out and made

[30] John D. Lee established a ranch on the Skutumpah Terrace in the fall of 1870 shortly before he was excommunicated by the Mormon Church for his alleged part in the Mountain Meadows massacre. With Levi Stewart of Kanab, Lee established a sawmill in Tenney Canyon, five miles from Skutumpah, but sold his interest in the mill in the spring of 1871 (Juanita Brooks, *John Doyle Lee: Zealot, Pioneer, Builder, Scapegoat* [Glendale, 1962], pp. 287–97).

Pang wich [Panguitch], 30 miles, a place of some 300 settlers—Mormons, of course. Pang wich means "fish" in Indian.

October 20, 1871

[Oct.] 20. Had a mule show and then started out. Just after leaving town saw a flock of sage hens. Bonemote shot two fat hens. Met a lot of Indians. Major traded two pipes for his tobacco. Pipes cut out of red clay. First I have seen of the Pai Ute make. They generally get them from the Snake Indians, who are very expert at pipe making. Went through a canon of the Severe. Rough road. Crossed the river about ten times, but only 12 miles in length. Went to a deserted town from which the Mormons had been driven by the Indians.[31] Called Circleville. A beautiful valley.

October 21, 1871

[Oct] 21. Started out. Passed through Marys Vale. Crossed the neck of a mountain. Reached Alma [Monroe] at 7 PM. Came 42 miles. Were entertained by Bishop [Moses] Gifford. Alma is situated in a beautiful valley, but every [illegible] is done by irrigation. Fine streams come from the Wasatch, clear as crystal and full of trout.

October 22, 1871

[Oct.] 22. Started out this morning early, and went to Glens Cove [Glenn's Cove, now Glenwood] to Bishop Neberker.[32] Got there about 10 AM. Had dinner and started out for Willow Creek [Mona]. Came 40 miles.

[31] Circleville was abandoned in June 1866 during the so-called Black Hawk War of 1865–68 (Peter Gottfredson, ed., *History of Indian Depredations in Utah* . . . [Salt Lake City, 1919], pp. 145–47, 176, 220; Gustive O. Larson, *Outline History of Utah and the Mormons 1776–1896* [Provo, Utah, 1958], pp. 169–70).

[32] The Latter-day Saints Church records show that Archibald W. Buchanan was Bishop at Glenn's Cove in 1871 (Marilyn Seifert, personal communication).

[Oct.] 23. Started early. Reached Manti by noon. Here we heard of the trouble at Salt Lake.[33] The Bishop was all in a flurry. Indians came in and told him that soldiers were on their way to arrest him. We left him in a state of boiling conditions. Started for Spring Town [Salem?] which place we reached about 5 PM. Came 28 miles.

[Oct.] 24. Pulled out early. Reached North Bend [Fairview] at 9 AM. Crossed the mountains to Spanish Fork, at which place we camped. Another surveying party came to camp with us. Had a pleasant time. This party, under command of a Mr. Fessem [?] intended to survey the Valley of the Severe.[34]

[Oct.] 25. Here Capt. Dodds left us for Uintah by the Spanish Fork Trail. Left the other party in the arms of Morpheus. Went to Springville, which we reached at 8:30. Found the stage had not come. We drove to Provo, the second largest city in Utah. Reached it at 9:30. Had come six miles. Here the Major took stage, while we fed and started for Leehigh [Lehi] where we camped. 39 miles.

[Oct.] 26. Started for Salt Lake, which place we reached at 2 PM, having come 35 miles.

[33] Hillers is probably referring here to attempts by the federal government, led by Judge James B. McKean, to suppress polygamy (Ray B. West, Jr., *Kingdom of the Saints: The Story of Brigham Young and the Mormons* [New York, 1957], pp. 316–24).

[34] It is not clear for whom this party was working. It may have been attached to the United States Geological Exploration of the Fortieth Parallel led by Clarence King (William H. Goetzmann, *Exploration and Empire: The Explorer and the Scientist in the Winning of the American West* [New York, 1966], pp. 430–66; Richard A. Bartlett, *Great Surveys of the American West* [Norman, Okla., 1962], pp. 123–215).

October 27, 1871

October 27. Stayed in the city 14 days. purchased 11 animals and two wagons. Brought Mrs. Thompson and Mrs. Powell back with us. Left the city on the 10. of Nov. Arrived at Johnsons Canon on the 19th.[35] Stayed one night—our party consisted of Major, Capt. Dodds, Bonny [George Otis] McEntee,[36] Mrs. Powell, Mrs. Thompson, Lavina Neberker [Nebeker], a young miss who joined us at Glens Cove,[37] and myself. Mr. Thompson met us about four miles from Johnson. Next morning I went to Kanab for flour and raisings. Found Steward, Jones and Beaman. Steward returned with me. Same day moved camp about two miles. Stayed two days.

November 22, 1871

22 Novem. Moved to Eight Mile Spring.

November 23, 1871

[Nov.] 23. Fixing camp. ~~Steward left for home on account of ill health.~~ [Crossed out in original.]

November 24, 1871

[Nov.] 24. Capt. Bishop, Fred, Clem, Andy and Ryley removed from House Rock.

November 25, 1871

[Nov.] 25. Still in camp hunting horses. Jones came down.

[35] Hillers's chronology is confused here. The section of the diary dated between November 19, 1871, and February 2, 1872, was apparently written on or after February 3, 1872, and Hillers remembered some dates incorrectly. Other diaries of the Powell expedition indicate that Powell and his party arrived at Johnson on November 30, 1871; the camp was moved to Eight Mile Spring on December 3. Bishop, Dellenbaugh, Clem Powell, Hattan, and Riley arrived from House Rock Valley on December 5, 1871. Thompson moved to a camp four miles south of Kanab on December 7 (Thompson, "Diary," pp. 62–64; W. C. Powell, "Journal," p. 372).

[36] Powell had hired George Otis ("Bonny") McEntee in Salt Lake City (W. C. Powell, "Journal," p. 372, n. 90).

[37] The Powells hired Lavina Nebeker as a nursemaid to help care for their new daughter.

[Nov.] 26. Prof., his wife, Jones, Bishop, Clem, Andy, McEntee left for Kanab Gap. Ryley and Capt. Dodds left for Kanab Wash, but camped with Prof. who retained them until the Major went down.

Moved our camp in Dec. below the Gap. from this point a series of observations was commenced, building monuments and measuring a base line ten miles long. Steward left us on the 4th of Dec.[38] Major, Dodds, Ryley and John Stewart[39] of Kanab went with the Major to find a way down to the Colorado by way of Kanab Wash.[40] Returned in about eight days time, having been successful in getting to the River. After his return he concluded to go East so of course I got everything in readiness.[41] Discharged Beaman. Major and him had a falling out.[42]

[38] John F. Steward had been ill during the latter stages of the river trip; he left to return to the East (John F. Steward, "Journal of John F. Steward," ed. William C. Darrah, *Utah Historical Quarterly* 16–17 [1948–49]: 250–51).

[39] Stewart was the son of Levi Stewart of Kanab.

[40] Powell and the men started from Thompson's camp on December 22, 1871, returning on January 2, 1872. (Thompson, "Diary," p. 64).

[41] Powell had decided to return to Washington to seek further appropriations for continuing the Survey. He was successful and returned to Kanab in early August, 1872. Thompson, meanwhile, remained in charge of the party.

[42] Thompson ("Diary," p. 67) records for January 31, 1872: "Went to Kanab with the Major. He settled up with Beaman." W. C. Powell ("Journal," p. 395) wrote on February 4, 1872, after he had reached Kanab, "We found that the Maj., wife, baby, Vina, and Jack had started to Salt Lake City on the way to Washington and that Beaman was discharged, the Maj. and Prof. being displeased with him." Beaman ("The Cañon of the Colorado, and the Moquis Pueblos," *Appleton's Journal* 11 [1874]: 548) puts things in a different light: "I with regret severed my connection with an undertaking that had my warmest sympathy." The archives of the Office of the Secretary of the Smithsonian Institution contain various receipts relating to the Powell expedition, among them one reading: "U.S. Topographical and Geological Survey of the Colorado River, J. W. Powell, A. H. Thompson. Jan. 31, 1872. For services rendered as Photographer from Apr. 1, 1871 to Jan. 31, 1872, 10 months, at 80 per mo. $800.00, Rec'd payment in full, [signed] E. O. Beaman."

Beaman went to Salt Lake City, procured a pack outfit, and made a trip to the Hopi mesas (see Beaman, "The Cañon of the Colorado").

1872

On the 2nd of February with four in hand drove Major and his wife and Lavina to Beaver. Camped at Pipe Spring with Bishop [Anson P.] Winsor.

February 2, 1872

[Feb.] 3d. Had an awful time finding horses — did not get started till about 12 AM. Camped five miles west of Short Creek.

February 3, 1872

[Feb.] 4th. Got off early and drove to Toquerville, stopped with Mr. Neberker.

February 4, 1872

[Feb.] 5. Snow and hail this morning, but nevertheless hitched up and drove to Kannarah [Kannaraville] 25 miles. I nearly froze. Stopped at the Bishops, but he not being at home nor any of the male portion the young ladies of the house had gone to a dance, nobody but the old lady being at home. Major, his wife and Lavina occupying a room of the parlour while I stretched my weary limbs on the floor of the parlor. Some time during the night I was slightly disturbed by the entrance of a young lady—everything being dark I could not see neither did I care. She soon began to call Jenny! Jenny, but I did not answer. She felt of my bed, and soon began to undress. I remained quiet and being fatigued and sleepy I soon

February 5, 1872

reembraced Morfeus in place of the young lady. On awaking next morning I found that my Bunky had left me but all her clothing and "other little unmentionables" I found by the side of the bed. None of the young ladies appeared at breakfast.

February 6, 1872

[Feb] 6. Hitched up and drove to Cedar [City] 18 miles where everything was arranged for them to take the stage, but when it came along the Major did not like to trust his family in such a rickety concern.

February 7, 1872

[Feb.] 7. Hitched up and drove to Red Creek [Paragonah] stopped at Bishop [Silas S.] Smith's—a small room was assigned to us, a partition dividing us from the "sitting room" or rather the nursery, as I shold judge from the amount of juvenile music that was made there. Going to supper we had to pass through this room, but what a sight met my eyes, though interesting to me as a bachelor, but "what a sight for a father." I counted 21 youngsters of age varying from 6 months to 13 years having passed in review. I next entered the kitchen and dining room being a combination of one. Here I was introduced to three Mrs. Smiths, none of them being over 32 nor yet good looking. The head of this vast multitude and strict follower of a passage in the Bible "Multiply and all increase man" being absent on business [at] Salt Lake.

February 8, 1872

[Feb.] 8. Hitched up and drove off for Beaver leaving the Smiths to enjoy themselves as best they could. Arrived at Beaver about 4 PM, having come 35 miles. Stopped at the Beaver House, kept by a Mr. Thompson, an apostate from the Mormon Church.[43]

[43] It is not clear to whom Hillers is referring here; he may have meant William Thompson who was in later years a federal marshal at Beaver (Juanita Brooks, personal communication).

[Feb.] 9. About noon I hitched up and started for home but did not go far—only about eight miles—Major took stage.

[Feb.] 10. Hitched up and drove to Red Creek to my old friends the Smiths whom I found still in a flourishing condition, but being unable to accommodate me with quarters, but being willing to allow my horses stabling, which was really all I cared for.

[Feb.] 11. Drove to Cedar to Bishop Lunn [Henry Lunt]. Took in 500 lbs. flour and 300 [lbs.?] beans.

[Feb.] 12. Started about 9 and drove to Dry Creek—camped out.

[Feb.] 13. Drove to Toquerville. Stopped at Mr. Neberkers.

[Feb.] 14. Hitched up and drove to the sheep troughs by way of Virgin City [Virgin, Utah]. Camped with Mr. [George?] Adair from Pa Rio Farms.

[Feb.] 15. Started early, arrived at Short Creek about one PM. Camped the rest of the day. Solitair and alone.

[Feb.] 16. Drove to Pipe Springs and camped at the Bishops.

[Feb.] 17. Drove to Kanab from Capt. Bishops all alone, the rest of the party having gone to the "Buckskins." [Kaibab Plateau]

| February 18, 1872 | [Feb.] 18. Resting. |
| | |

February 18, 1872

[Feb.] 18. Resting.

February 19, 1872

[Feb.] 19. About noon Prof. and his party returned. Packed up and drove to Kanab Gap and camped. Alfred Young joined party as herder.[44]

February 20, 1872

[Feb.] 20. This morning I was installed as assistant photographer. Clem tried to take some pictures but failed, bath being out of order—fixed it—while manipulating he upset it—so much for the first day.[45]

February 21, 1872

[Feb.] 21. Clem went to town for Beaman's old bath—fixed it up by filteration.

February 22, 1872

[Feb.] 22. Put up the dark tent—Clem put the bath in it open, started to a spring for some water about a quarter of a mile from camp. On our return, lo, our tent had blown over and the bath spilled again. Tried another which he upset a third time. While coming out he mixed up another. George Adair joined us as packer.[46]

February 23, 1872

[Feb.] 23. Tryed our bath which worked all right. Packed up and started for Stewart Ranch. Camped in Oak Canon by a small spring.

[44] Thompson ("Diary," p. 69) calls him "Alfred Zenng"; Jones ("Journal," p. 111) calls him "Alfred Young."

[45] W. C. Powell had been Beaman's assistant throughout the river trip. He had much trouble with the complicated wet-plate photographic process. Thompson ("Diary," p. 68) attempted to help him but with little success. Clem was eventually replaced by James Fennemore and later, by Hillers.

[46] "George Adair commenced work at $40.00 per month" (Thompson, ibid.).

[Feb.] 24. Packed up and started for the ranch. Began to snow which lasted all day. Got in about noon. A beautiful spring [Big Spring] starts from the side of the mountain about 250 [ft.] up and comes down on a slope of about 80 degrees. Water enough to run a turbin[e] wheel. Some enterprising man besides Bishop Stewart ought to own it. Lumber could be sawed here to fence in all Utah. Magnificent. Pines measuring 125 feet in hight. Yellow pine.

February 24, 1872

[Feb.] 25. Still snowing. In camp. Clem and myself started down. Got three views.

February 25, 1872

[Feb.] 26. Tryed to take views but snow drove us in after taking two.

February 26, 1872

[Feb.] 27. Clem and myself started up the canon. Took three fine views. Clem took two while I took one.

February 27, 1872

[Feb.] 28. Hunted horses which had wandered off during the storm. Failed to find them.

February 28, 1872

[Feb.] 29. Hunted horses again. Found them. Jones and Dodd started to find a way to the river across the mountain.[47] Andy and Fred started to the southwest point to build a monument.

February 29, 1872

March 1. Dodds and Jones returned. Snow prevented them from going on. Some places 10 and 15 feet deep.

March 1, 1872

---

[47] "Captain Dodds and Jones started to go as far on Kaibab as they could" (Thompson, "Diary," p. 69).

March 2, 1872

[Mar.] 2. Polished glass.[48]

March 3, 1872

[Mar.] 3. Clem and myself started with eight days' rations down the Kanab Wash. Took two pictures. Camped at the mouth of Oak Canon.[49]

March 4, 1872

[Mar.] 4. Packed up and started about 10 AM. Arrived at the Wash about 4 PM. Camped at some water pockets.

March 5, 1872

[Mar.] 5. Started early. Made the head of Running Water about 5 PM and camped. Canon walls begin at zero from the pockets and run up to 3000 ft. Almost perpendicular. Limestone formation.

March 6, 1872

[Mar.] 6. Rain this morning. Started to go to the River, being only about six miles from it, Bonny and Ryley being down there, washing gold, having left the party while I was off with the Major. But after having gone only about two miles it cleared up, so we concluded to go back, but on returning it began to rain again.

March 7, 1872

[Mar.] 7. Started back. Photographed all the best scenery. Camped at the head of the water.

March 8, 1872

[Mar.] 8. Packed up and off early. Took views wherever we could get them. Camped near a pocket in the Rock.

[48] That is, glass for photographic negatives.

[49] W. C. Powell's diary for this period is missing. A letter by him published in the Chicago *Tribune* July 11, 1872, contains only this note: "On the 3rd of March I started out, with Assistant Jolly Jack to take views."

[Mar.] 9. Filtered our bath and started taking views as we went. Got to the Water Pockets about 7 PM, but oh, horror, our Pockets were empty—the rest of our party had camped there and emptied them—no water for us that night.

March 9, 1872

[Mar.] 10. Up early hunting water. My search was rewarded by finding a large pocket full—what a Godsend it was. A man can stand hunger but not thirst—he gets fairly crazy. Had breakfast and started for Pipe Spring. Found some fine scenery—had to stop—took four views. Camped at the head of Running Water. Met Mr. Winsor and six miners. Winsor hunting horses while the miners were on their way to the River after gold.

March 10, 1872

[Mar.] 11. Started early—got into Pipe Springs about two. Found mail for me from San Francisco and all the "Agles", Also William Johnson [William Derby Johnson, Jr.,] of Kanab who joined the party as assistant topographer.[50]

March 11, 1872

[Mar.] 12. Commenced to fix a new Dark Tent.

March 12, 1872

[Mar.] 13. Working at Dark Tent. The Deacon [Jones] and Johnson started for Wolf Spring to take bearings.

March 13, 1872

[Mar.] 14. Finished my Tent. Deacon and Johnson returned.[51]

March 14, 1872

[Mar.] 15. Commenced a Packsaddle.

March 15, 1872

[50] Johnson was a school teacher in Kanab and had become well acquainted with F. M. Bishop (Darrah, *Powell of the Colorado*, p. 449).

[51] Thompson ("Diary," p. 71) records Jones and Johnson as leaving for "Signal Station" on March 12, 1872 and returning the following day.

March 16, 1872	[Mar.] 16. Finished my saddle.
March 17, 1872	[Mar.] 17. Made Sinches [cinches].
March 18, 1872	[Mar.] 18. Tryed our new Tent which proved a success.
March 19, 1872	[Mar.] 19. Packed up for a trip for the Uinkarets Mountain. About noon a Mr. Fenimore [James Fennemore] reported to Prof. as Photographer, having been engaged by the Major at Salt Lake.[52]
March 20, 1872	[Mar.] 20. Clem felt blue because he had to turn the instruments over to Fen. Told me that he would leave party as soon as his year expired. Fen. took picture of "Winsor Castle." [53]
March 21, 1872	[Mar.] 21. Up and off early. When about half mile from Pipe Springs the pack on our mule became loosened. I called to Clem to assist me. While in the act of fixing it his horse scampered off, following the others. I told him to take my mule and follow. He soon returned with the horse, but his gun which was fastened to knob of saddle had been lost off. Clem remained behind to look for it.[54] Prof. and Fenimore started also back, but their search was in vain. They soon returned leaving Clem looking for it. Went about 14 miles and camped near a pocket. Saw three wild horses. Up to late—Clem had not come.

[52] Fennemore was an assistant in the Savage and Ottinger gallery in Salt Lake City. Powell convinced him to join the survey; he remained until mid-August, 1872.

[53] The fort at Pipe Springs, constructed under the supervision of Bishop Anson P. Winsor (see W. C. Powell, "Journal," p. 400).

[54] Clem valued the rifle, a Winchester, highly. It was given to him by the Major. He found it after a day and a half search (W. C. Powell, "Journal," p. 402).

100

[Mar.] 22. Packed up and off early—reached the foot of the mountain [Mt. Trumbull] about 5 PM. Hunted for water but could fine none. I pitied the poor horses. No coffee.

March 22, 1872

[Mar.] 23. Struck an Indian trail. Packed up an[d] off before breakfast—followed it for about five miles which led to a large pocket full of water. The joy I felt I cannot express. Horses drank heartily. We soon prepared breakfast. Camped for the rest of the day. Fen and myself made views of the Pocket and Gulch.

March 23, 1872

[Mar.] 24. Sunday being in camp, but Fen. took picture of the mountain. Called it "Mount Trumbull." [55] The Deacon and Fred went out to look for a Ranch [Whitemore Ranch] which was supposed to be somewhere about this mount —they fortunately found two men in search for stock from the Ranch. Prof. and Dodds went on the East side to look for water about 20 miles down. Fen and Johnson went to look for fossils. Mrs. Thompson gathered plants. Prof. and Cap. returned about 7 PM, having found water and also the Ranch.

March 24, 1872

[Mar.] 25. Up and off early for the top of the mount[ain]. Prof. and his wife, Jones, Dodds, Johnson, Fred., Fen. and myself built a monument. After that all but Jones, Fred, Fen. and myself returned while Fred and the Deacon took bearings, while Fen. and myself tryed to photograph but it was too hazie. Stayed all night.

March 25, 1872

[Mar.] 26. Begun to take views until noon, then went to the south end and descended, camp having been removed in that direction. After traveling about an hour and a half we discovered camp in an oak grove with a spring in the center.

March 26, 1872

[55] Named for Lyman Trumbull, U.S. Senator from Illinois, 1854–73.

March 27, 1872

[Mar.] 27. Jones and Dodds started to find a way into the River.[56] Fen. and myself took three views of a lava bed—a recent outpouring. Returned at noon.

March 28, 1872

[Mar.] 28. Took three views about camp. In the afternoon started toward the Colorado, had an awful time getting down to the foot of the mountains, it being all cut up with gulches and the lava being cut up into large ragged blocks. Took three views. Returned late at night. Could see the Canon of the Colorado. Saw smoke near the foot of a black Hill near the Canon.

March 29, 1872

[Mar.] 29. Packed up our mules and started back to where we passed down to the valley. Took two pictures. Found some Shenamo[57] hyrogliphics—tryed to copy them with the camera—failed on account of a storm coming up. Returned to camp about noon. Found Jones and Capt. Dodds home. Found that the smoke which I had seen yesterday had been their camp fire.

March 30, 1872

[Mar.] 30. Jones and Fen. left for Kanab to bring rations and chemicals for photographing. Prof., Capt. Dodds and Johnson started to hunt a place to get down to the Colorado.

March 31, 1872

[Mar.] 31. Snowing and raining alternately. Prof. returned this evening.

[56] The men were searching for access routes through which to bring supplies for the projected river trip into Grand Canyon during the coming summer.

[57] Powell used the term "Shinumo" to refer to the Hopi Indians, as well as to the Puebloid archeological sites found in the country. Hillers is here following his usage. The term "Moqui" was also used synonymously (see April 1, 1872, entry).

April

[April] 1. Andy and myself started toward the Colorado—took a picture of some Moqui hieroglyphics which were situated in a gulch near the foot of Mount Trumbull. Rode down to the River—got there about 5 PM. What a sight met my eyes! Looking down on the river from the top it appears to be nothing but a narrow gutter from the top to bottom—I should judge about 4000 feet and about 400 feet wide. It looked gloomy and forebidden. I counted eight rapids whose roar came up like a distant roll of thunder. Got to camp—found Andy waiting with supper. Camped by the side of a pocket.

April 1, 1872

[April] 2. Rain this morning. Could not make pictures.

April 2, 1872

[April] 3. Still raining. Our ration gave out. Took two views during the storm. Returned and got to camp about 3 PM. Found Capt. and Johnson had returned about sundown, having been successful in finding a road down to the River.

April 3, 1872

[April] 4. Snowing this morning. Fixed up to go down again but Prof. concluded not to being short of rations.

April 4, 1872

[April] 5. Packed up and started for Pine Valley. Snow in some places two feet. Lost the trail and started in a straight line for Berry's Spring. Made dry camp.

April 5, 1872

April 6, 1872

[April] 6. Started early but at ten it began to storm. Came to the edge of a line of cliffs—could not see the valley below. Unpacked and camped till one PM. Packed up and off—snowing at intervals. Made another dry camp.

April 7, 1872

[April] 7. Up and off early. Followed the line of cliffs all day. Finally came to a gulch which cut us off—perpendicular walls on both sides—could not find a place to get down in the valley. Camped near the gulch plenty of water. All we had was bread and coffee for supper, being our last.

April 8, 1872

[April] 8. Breakfast consisted of a pot of beans boiled in water no meat. Put out and struck for Gould's Ranch. Purchased some corn meal and molasses, then put out for Berry, where to our great joy we found our old friends Jones and Fenemore, also Clem and George Adair with lots of grub. My shot gun was also among the rations.

April 9, 1872

[April] 9. Found two letters and lots of newspapers. Commenced to fix a new dark tent.

April 10, 1872

[April] 10. Prof. and Adair started for Toquerville.

April 11, 1872

[April] 11. Fixed a dark tent and went hunting ducks. Shot three.

April 12, 1872

[April] 12. Packed up for a trip to the Uing karrets [Uinkarets] again, Capt. Dodds, Frederick, Fennemore and myself, Dodds to find a way to the river at several points, Fred to finish triangulating which could not be done during the storm and rations would not hold out till it might clear up, while Fennemore and

104

myself went for pictures, which I failed to get on account of the weather. Camped the first night at Fort Pierce [Fort Pearce], a place used as a defense against the Indians but now abandoned.

[April] 13. Struck out early, took the wrong trail and went four miles out of our road, but finally found the trail. Kept along the cliffs. Camped near a pocket which we fortunately found.

April 13, 1872

[April] 14. Up and off early, still following the [Hurricane] cliffs. Saw four antelopes but too far to shoot. Dry camp tonight for the horses but plenty feed.

April 14, 1872

[April] 15. Still following the cliffs which go clear to the mountain. These hills, or rather the plateau on top, is called Hurricane Hill. Climbed the mountain and went to the east side to our oak grove camp, where man and beast drank heartily.

April 15, 1872

[April] 16. Fennemore and myself packed up and started for the river. Camped near the pocket where Hatten and myself had camped before. Day being dark and cloudy could not make pictures.

April 16, 1872

[April] 17. Started for the river. Tryed our luck but failed. Bath out of order. Started back and mixed up a new one.

April 17, 1872

[April] 18. Started down again, and made some fine views of the river.

April 18, 1872

[April] 19. For the river again. Obtained seven views. Fred and Capt. came down, having finished their work. Capt.'s Indian did not come to pilot him.

April 19, 1872

[April] 20. Capt. and Fred climbed down to the river. Fen. and myself started for more pictures. Made ten and returned to camp near pocket.

[April] 21. Packed up and started for our old camp in oak grove. Arrived about 3 PM.

[April] 22. Up and off early. Fred started for the ranch kept by Whit[e]more of St. George, in order to map the trail from the ranch to St. George. The rest started down by way of our old trail. After riding about four hours we met a man driving a mule loose while he himself carried the saddle, mule having broke his hobbles during the night, and was unable to catch him. We offered him help but the fool declined it. Asked him about the water, but his descriptive powers being on a equal with those of the "intelligent" contraband of the south, we could not find it where he said it was, or at least where we understood him to mean, but found a small seep in a gulch in the cliffs, the coyotes having dug it out, but it failed to supply us with the requisite amount, but concluded to stop and wait for Fred, but on retiring to bed had not come. I am very uneasy concerning his safety.

[April] 23. Up and off. Fen. and myself with the pack train. Exchanged animals with Capt. Dodds, my mule being a better traveler than his mare, him to go back to the ranch to look for him, we being off the trail about two miles, but on reaching the trail I found that he had passed—saw his mule tracks. I felt happy but sorry for Capt, but still being in hope that he would follow the trail on which we had gone to the mountain. We started for a pocket 21 miles from our last camp at Coyote Springs, but on reaching this pocket all the water had left. I pittied poor Fred—no water since yesterday morning and no grub. He left a note that he had gone to the next pocket. Of course our animals, nearly perishing for water, had to go 12 miles more, and that might be dry too, for all we knew, for this pocket was full when we went up, but now not a drop in it. But on reaching it it still had plenty of water and the way the poor beast drank it I thought it would burst them. Found Fred had gone for corn 26 miles from here. Capt. returned at 11:45 PM, having found Fred's tracks after going ten miles, having traveled 55 miles in ten hours, and a hot day. Horses have no business in this country. Mules and jackasses are the only animals that can stand up in this dry country.

[April] 24. Started for camp by way of Fort Pierce. Stopped at the Fort to water and bait our horses and had a lunch of bread, coffee and jerk. Arrived here at Berry's Springs at 4 PM. Found mail from girls and [illegible] also a letter from John Doyle [Lee?].

April 24, 1872

[April] 25. Duck hunting this morning but got only one. All I saw. Wrote letters.

April 25, 1872

[April] 26. Shot two ducks, finished my letters. Jones went to Toquerville with the mail. Cleaned glass and fixed up the traps for a photographic tour to Mt. Kolob.

April 26, 1872

[April] 27. Broke up camp. Prof. and wife, Fred, Adair for Mary's Mt.,[58] Jones and Johnson for the Pine Valley Mts., Capt., Andy and Alf. for Kanab, Fennemore and myself for Kolob [Plateau]. Left about noon by way of Hurricane

April 27, 1872

[58] Thompson ("Diary," p. 76) includes W. C. Powell in the party and indicates its destination as Sharp Mountain (Mount Bangs).

Hill and camped for the night at bank of the Virgin about half mile below town of that name [Virgin City]. In the evening we visited the bishop of the place, a Mr. [John] Parker. What a relief to the eye to look upon an orchard all in bloom after being out in the sage brush for a number of months. He seems to be lost, he is fairly enchanted. Scenes of childhood, long forgotten flash up before the eye of memory in vivid succession—how when a boy he looked forward to the time when those trees, now in bloom, should be heavy laden with fruit—how he would watch day after day until fit for digestion. We spent a few hours with the Bishop and his wife in pleasant conversation. Nice folks they. Returned to camp, and consigned ourselves to Morpheus.

April 28, 1872

[April] 28. Packed our mules and started for the Mt. Called in at the Bishop's whom we had promised to make a picture of his house, but on account of the high wind and clouded state of the weather, we were unable to fulfil our promise. Procured a canteen of wine, the sample last night being very good. Off we started for the top, following a trail which led up a creek very rapid and shallow, called by the natives North Creek. All along the bottom for seven miles where a small spot free from rock appeared, it was under cultivation, mostly vineyard. The grape seems to flourish here. This southern Utah is called Dixie. Near the head is a saw mill. Lumber is made here for all those who haul the logs, which they get from the top of Kolob, a distance of ten miles. Having climbed some 3000 feet we encountered a snow storm and my old soldier came into requisition. What a change from four hours ago, people walking about in their shirty sleeves. The scenery is grand of its kind and affords fine subjects for the camera. The formation is homogenus sandstone. The top had been overflowed with lava. Some parts of a snow white, while others of it a dark red and intermediate color. Pinnacles innumerable, forming an immense harrow upside down. Here and there are scattered peaks, overlooking the others by a thousand feet, or like giants among dwarfs. All bare rock, no vegetation on the towers. Camped near a little seep spring.

[April] 29. Up and off for views; after going about three miles we came to a house half finished and a beautiful spring. Stock is kept here during the summer. We unpacked and tryed to make pictures but failed on account of the high wind and storm of snow. Camped in the house. Toward evening Fen and myself took a walk toward the head of a canon, climbed down about a thousand feet through brush and rocks. Found a big spring running out from under a bed of lava. Preferring to climb up the vertical wall of lava to scrambling back through the brush we began our task, but before getting half way up we felt sorry for our bargain. Reached the house quite late. Was surprised to see a big fire in the house, but on nearing found a visitor in the person of a Joseph Gibbs, of Dunken's Retreat,[59] in search of stock.

[April] 30. Out early on the hill. Made ten pictures, then returned to dinner, bringing with us our traps. Had an awful time catching our animals. After dinner packed up and started to the Twin Peaks of pure white sandstone. Made two views and returned. Our visitor had left us, but what a fright—the bars which we had put up in front of the door had been removed, but on nearing found that they had been thrown down by some ferocious brute. All our cooking utensils was buried in the mud and sand. A pound of fresh butter which we had left on a plate could not be found at all—probably eaten for the salt. One spoon and fork could not be found.

[59] A farm between Toquerville and Rockville established in 1861 by Chapman and Homer Duncan (Juanita Brooks, personal communication).

Our negatives and chemicals were arranged along the wall which miraculously had escaped his wrath. Probably the smell prevented him from molesting them.

May 1, 1872	[May] 1. Packed up and started for an Alcove Cliff. Took three views, then returned to our first water. Made three views and camped, hungry and weary from our day's work.
May 2, 1872	[May] 2. Packed up and started for Virgin City.
May 5, 1872	May 5. Washing glass. Jones and Johnson joined us.[60]
May 6, 1872	" 6. Polishing glass. No party.
May 7, 1872	" 7. Cleaning glass.
May 8, 1872	" 8. Washed my clothes.
May 9, 1872	" 9. Party returned at 9:30 PM.[61]
May 10, 1872	" 10. Packed up and started for Kanab. Camped just below town.

[60] There are no diary entries for May 3–4, 1872. According to Jones's "Journal" (p. 124), Hillers and Fennemore arrived at Pipe Springs on May 4. Jones and Johnson arrived there the following day.

[61] The "party" consisted of Thompson, his wife, W. C. Powell, Dellenbaugh, and George Adair. They had made a reconnaissance into the Virgin Mountains southwest of St. George and had returned to Pipe Springs via Fort Pearce (Thompson, "Diary," pp. 76–77; Jones, "Journal," p. 125).

" 11. Had to cook, that functionary [Hattan] having gone to the mouth of Paria. Fenimore fixed up for printing pictures.

[May] 12. Stopped cooking and went to work with Fenimore.

[May] 13. Jones took a team and started for Pipe [Springs] to bring the rest of our stores.

[May] 14. Still printing. Jones, Andy and Capt. returned this evening.[62] Weather cold and rainy.

[May] 15. Printing finished this day.

" 16. Mounted our pictures.

" 17. Still to work mounting.

" 18. Fixed up to start for Dirty Devil.[63]

" 19. In camp fixing up.

" 20. Still fixing.

May 11, 1872

May 12, 1872

May 13, 1872

May 14, 1872

May 15, 1872

May 16, 1872

May 17, 1872

May 18, 1872

May 19, 1872

May 20, 1872

[62] Hattan and Dodds had been to the mouth of the Paria River to check on the boats the party had cached there the previous fall (Thompson, "Diary," pp. 76–77).

[63] Thompson was preparing to start on an expedition to locate the mouth of the Dirty Devil River and to retrieve the *Cañonita* which had been cached there in the previous year. The party consisted of Thompson, Dodds, Hillers, Dellenbaugh, Fennemore, Jones, W. C. Powell, Hattan, Johnson, and Adair.

May 21, 1872	" 21. Made Clem a dark tent, to[ok] a feriotype of Fred.
May 22, 1872	" 22. Still in camp.
May 23, 1872	" 23. Everything ready for tomorrow's start.
May 24, 1872	" 24. Prof. decided to wait till tomorrow.
May 25, 1872	" 25. Off this morning Eight Mile Spring. Found no water. Off for Johnson Canon. Camped about 2-½ miles above settlement.
May 26, 1872	[May] 26. Sunday. Taking it cool
May 27, 1872	" 27. Hitched up and started for Kanab with the wagon, returned same day, bringing with me George and Fen. who had remained behind. Fen. forgot his filtering tank, returned for it but failed to catch up. No sign of him up to ten PM.
May 28, 1872	[May] 28. Fenemore came to camp about noon, having got lost during night. His mule broke away and so had to foot it. A party of prospectors came from Salt Lake. Camped with us. Willie Johnson paid us a visit.
May 29, 1872	[May] 29. Fenemore went back after his saddle—returned at noon. Prof. came up from Kanab about 9 AM. Prof., Fred and Johnson went up on point "B" for observations. Returned at night bringing Fen's mule which they found with a band of Johnson's horses. Alfred Young started for Kanab. Struck for more wages which was refused and so Prof. discharged him.

[May] 30. Got our pack ready, started about 10 AM. up the canon, reached Lee's Ranch about 4 PM.[64] I understand from George that J.D.L. has sold out for 1800 [dollars] to a man by the name of [John Wesley] Clark. One of Lee's women [Lavina Young Lee] still resides here, while he has another with him at the Paria.[65] This is a splendid place for stock. Fred and myself climbed up a little butte. Fred sketched the surrounding cliffs and valley.

[May] 31. Started early, struck across the country in a northerly direction. Towards evening struck a nice little valley [Adair Valley] with a clear cold spring [Adair Spring]. Concluded to camp. After supper went out to hunt. At the lower end of the valley found a beautiful lake. Saw lots of ducks. Killed three. The rest hiding in the reeds which surround the lake. Adair put his preemption stake at the spring. Called it Swallow's Park, and Lake Adair. Indian guided us across.[66]

May 1 [June 1.] Off at an early hour. Struck an old trail which is supposed to be the trail of a hundred men who started from Kanab in '67 [1866] on an exploration of the country in general, all Mormons, a man by the name of Jim Andrews [James Andrus] in charge. After winding and twisting up and down hill for 16 miles we went down a large gulch containing running water. Our attention was attracted by two tombstones but the grave had been dug up either by Indians or wolves as the bones were scattered about. This grave it is supposed contained the

[64] See diary note 30.

[65] Emma Bachellor Lee was established at Lonely Dell (later Lee's Ferry) at the mouth of the Paria River; Rachel Woolsey Lee was established at Jacob's Pool in the House Rock Valley (Brooks, *John Doyle Lee*, pp. 303–316).

[66] "Indian Tom," a Kaibab Paiute, had agreed at Skutumpah to guide the party to Potato Valley (Jones, "Journal," p. 128).

body of Everett, [Elijah Averett] one of the 100. This Everett and seven others were sent back with fatigued horses to Kanab. While going up the hill Everett was first, a couple of Indians having tracked the party, seeing these return, fired on them and killing Everett. The rest being none but boys fled back, the Indians carrying off their horses. Had Everett followed the instructions of Andrews his life would have been saved. Among the band of horses which they took back was a mare who could smell Indians a quarter mile off and could not be got to go up to one. The instructions from Andrews were to drive this mare ahead to make sure, but they led her in place of driving her. One of the boys returned to camp, told his story, when the party gave chase recapturing the horses and guns. The Indians fled to the mountains, which afterward were captured and roasted. Camped at Pa Rio [Paria] River.

June 2, 1872

[June] 2. Went up the River to its head.[67] When about 12 miles from camp I recollected that I had forgotten my ammunition. I returned, found it all right—was overtaken by a regular mountain thunder shower which saturated every stitch on me. I followed the track of the party which I found had gone over the divide into Potatoe Valley. Climbed up on a narrow strip resembling a hog back, from bottom to top 2500 feet—in some places it was only 2-½ feet wide on top. The sun was just going down. When I reached the top I stopped to breathe. I looked down the awful chasm—it looked wild and forbidden, the valley below being in deep shadow. Table Mountain [Table Cliffs] north of it with its pink colors reflected back the golden light of the setting sun long after she had gone from my view. A wilder view I never beheld. It surpassed the Wild Scene on the Colorado at Lava

[67] That is, the head of Henrieville Creek (Thompson, "Diary," p. 80).

Falls. What a picture for Burstadt [Bierstadt][68] After leaving the edge of the cliff I found myself in a beautiful little valley hemmed in by mountains on each side. Green grass dotted with flowers—in the distance I saw the blazing camp fire of my companions who probably feared I might not reach it, as night set in very dark, but my faithful little mule never missed a step from the trail, having come 42 miles. Supper was eaten with a relish which only a mountaineer enjoys. After supper a smoke of "Navy", after which I joined the chorus of snorers.

[June] 3. Our Indian felt sick and wanted to go back. Prof. gave him an old horse blanket and some flour then "Pike way'd" for home or some other place. We followed down the valley. This is the head of the Dirty Devil River[69] it starts by springs. At night found a large creek [Escalante Creek] coming in on the left, where we camped.

June 3, 1872

[June] 4. Rained all night and all day long. Prof. feeling sick concluded to rest.

June 4, 1872

[June] 5. Packed up and started down the valley about 8 miles. Here the river ran into a narrow canon which made us halt. Unpacked—had dinner. A tilted ledge about a thousand feet high runs square across the valley running northwest

June 5, 1872

[68] Albert Bierstadt (1830–1902) who was famous in the nineteenth century for his grandiose landscapes of the Rocky Mountains and the West (John C. Ewers, *Artists of the Old West* [Garden City, 1965], pp. 174–85).

[69] Upper Potato Valley is not the headwaters of the Dirty Devil River, but a part of the drainage of the Escalante River as Thompson ("Diary," p. 81) concluded three days later (see below).

and southeast a quarter of a mile in the canon—a rapid little creek comes in on the left, coming down from Lake Mountain [Acquarius Plateau] who loomes up in the distance with his snowy garb and tapering pines. After dinner everybody climbed out or looked for a short cut to the mouth of the Dirty Devil River. Fen. and myself climbed up the ledge with instruments to make pictures—succeeded in getting two fine ones showing the junction of the two streams and also the Canon. Returned to camp. The General soon after spread the cloth for supper. Capt. and Prof. made their appearance on the hill and soon joined us at supper. The[y] reported having followed a trail twelve miles leading to a gulch with plenty of water pockets.

June 6, 1872

[June] 6. Packed up while Prof. and Capt. went off on the trail trying to follow it. I brought the train to the gulch [Harris Wash] and camped. Fen. and myself started down the gulch for some pictures. After going about a mile we came to a jump down of nearly a thousand feet deep, while the walls above us nearly measured the same. It gradually tapered down making it look like a huge bin, top of Canon being nearly 2000 feet wide while at the bottom on[ly] six feet, no water being in it, only in pocket. Photographed it and two other views, and then returned to camp. Found Prof. and Capt. had returned. Prof. reported that he was satisfied that this creek which we had been following was not the head of Dirty Devil River—what could it be—but finally came to think that a small stream came in on the right just above the San Juan River—being now about only 50 miles from Colorado River, for Mt. Seneca Howland [Navajo Mountain] being in plain sight, and the whole country seemed to sink towards it, while the Dirty Devil Mountains [Henry Mountains] loomed up 75 miles north of us and the river flowed on the northern side, so it was impossible for this stream to take a turn and flow up grade.

[June] 7. Prof., Capt., Clem and Adair went to a high point to obtain a panoramic photograph of the country while I took packtrain back to our old camp which we had left yesterday morning. Fen. and myself, with animal started down the Canon a little ways for pictures but could not get down a great ways, for where the two creeks join their water(s) form a deep stream, especially this season of the year. The creek coming down from the west is called Lake Creek and the other Birch [Creek]. Made three pictures then returned. Found the rest of the party had also arrived. Soon spread the cloth and "went for it," I mean supper. Prof. had concluded to send three men [George Adair, S.V. Jones, and W. C. Powell] back to Kanab for Rations and return to this place as soon as possible. I was designated to bring the boat down, also Fred, Johnson and Fennemore. I hastily scribbled a few lines to my friend Dick Geits.

[June] 8. Adair, Clem and the Deacon started off for Kanab with seven animals, while we packed nine and rode seven. Started up Lake Creek toward Lake Mountain but left the creek near mountain to our right where the canon followed a Gulch at the foothills for about six miles. Stopped for ¾ of a hour. Prof. and Capt. climbed out but soon returned and reported that we had to climb up—up we went 2000 feet—a grand sight is had from the top. About sundown we passed around a mountain—the scene was grand—it put me in mind of crossing the Isthmus of Panama. Struck a beautiful little Creek [Boulder Creek?] which came rushing taring down from the top of the crest. Its sparkling cold water was enjoyed by both men and beast. Feed the best I have had since leaving the States. Camped for the night.

June 9, 1872

[June] 9. Packed up and off. This mountain travel is fearful—no sooner have you reached the top of some Giant than you will have to climb down again on the other side some thousands of feet. Reaching the bottom or rather the divide you find a canon which cuts you off—now you have to go up to its head. Perhaps you are a lucky Dog and find where the water has undermined the bank and let it slope down at an angle of 60°. Now put on the corsetts, but no matter how tight you make them something will give way and away goes your pack. You're lucky if you find everything together again—some is sure to reach the water and then that is the last of it, for the water rushes down with a tremendous force and sweeps it beyond your reach. Followed up quite a stream for about five miles—it ran down at an angle of 30°. It came down leaping, taring, rushing, now running down a smooth slope, now taring over big boulders, now leaping down 50 or 100 feet at the bottom of which it finds itself churned into white foam—here it collects itself and again is off, never stops until it finds itself in the Pacific Ocean, via Colorado River. We crossed seven beautiful streams—all along these streams the finest pasture I saw in Utah. Winding around a huge old Mt. all of a sudden we saw a beautiful lake [Aspen Lake] about half mile wide and the same in length. Fenemore and myself made a picture of it. After supper went down and shot several ducks.

June 10, 1872

[June] 10. Poor Fen. is sick this morning with cramps in the stomach, but he rode off with us. Stopped at noon by a roaring stream. Before reaching noon camp Capt. climbed up a tree and reported two lakes east from our trail, also reported a beautiful valley, for which we started and nooned. Prof. concluded to remain here, Fenemore feeling too sick to travel—Fred and Johnson went with me to the lakes, which I photographed. Called the "Bee" Lake. Fred called the other Hidden Lake. The finest place for a ranch I have seen in Utah is right here. This valley including some low hills, that is, the foot hills of the Lake Mountains, is 20 miles square, with streams running down every five miles. Feasted on Pine hens.

[June] 11. Long time finding our horses. Did not get started till about 9:30. After filling themselves with the long green grass they had taken to travelling; about two miles from camp we struck a stunning creek—it ran the most water of any we have seen—it ran very swift, jumping and foaming as it went down the steep slope, full of little falls called it Cascade Creek from the top of the mt. It leaps over a perpendicular cliff over a thousand feet high, but its fall is broken by a large rock sticking out from the side about three hundred feet from the top—the rest of the way it goes a little slanting at an angle of about 80°. Did not go close to it for our time is precious, and rations short. Crossed in all nine creeks of different sizes, all running into Burch Creek and then into the Colorado. In the afternoon we left the southern slope of the Lake Mountain Range and crossed to the east side. The country below us is all cut up with gulches and canons for miles—nothing but sand rock is visible. The Dirty Devil Range [Henry Mountains] from this point looks like a dry country and almost impossible to get to them. The first peak is about 30 miles from this point eastward.

[June] 12. Prof. and Capt. went in a northeasterly direction to find a trail or hunt up the best place for to get to the mountains near which our river flows. Fred and Johnson went southeast on a similar errand. Fen. and myself went down a canon due east for pictures. Secured three—a storm coming up we returned to camp. Shot a dusky grouse; on our return found Fred and Johnson home. They reported having found a heavy beaten but not very old trail leading in the direction of the mts. About 5 PM Prof. and Capt. returned. Reported having seen fresh Indian sign. Rained very hard.

June 11, 1872

June 12, 1872

June 13, 1872

[June] 13. Followed the trail F. and J. had found which led us through small valleys and canon and to wind up, lead us down from the plateau on smooth sand rock over a thousand feet high—horses often slipped for a number of feet. Sometimes found ourselves on projected shelves, not over two and a half feet in width, but got down all OK. The last turn brought us in sight of a beautiful valley running northeast [Pleasant Creek] while from the west came a rushing stream which flowed in a small canon of sand. All along the edges the cottonwood trees flourished. Camped in the canon. Plenty feed for our animals—wild oats in abundance. Followed fresh Indian tracks all day.

June 14, 1872

[June] 14. Up and off early. While making a descent from a Bench we were attracted by the bark of a dog. Looking in the direction from hence we heard it we saw two squaws flighing through the grease wood, yelling as though they had seen so many devils. We saw their camp on a small hill for which we started. On nearing an old man about seventy met us at the foot of the hill, trembling like an aspen leaf. On reaching this camp found it deserted. Guns, bows and arrows had been hastily left in their wickiups, but we soon dismounted and seated ourselves around his camp fire, where we allayed his fears by telling him to smoke, at the same time handing him tobacco. At this token of friendship he steadied his nerves and began to talk. Found him to be one of the Red Lake Utes down here to gather seeds. A quarter of a hour after in came the two frightened squaws, who began to chatter like monkees. [A] little while after we saw two young men across a gulch on a small hill. The Squaws began to call to them not to be afraid for we were Tuitchea Ticabu (very friendly). They soon came slowly toward us looking like shamed men—no doubt they felt so. After smoking and talking, They begged us to go into camp and trade with them. Prof. being anxious to know about the country at large, and all about the trail, he consented to camp—traded several buck skins. I obtained a splendid skin for a small paper of paint.

[June] 15. Struck off on a trail up the valley—followed it into a canon. About a mile from its mouth we lost sight of it with the exception of a track here and there —probably cattle feeding. This is the great hiding [place] of the Indians and many heads have found their way in here, all stolen from the Mormons, who never suspected for a moment that their friends the Utes would do the like, but thinking the Navajoes the guilty Party. The Utes always watched the opportunity, when a band of the former were in the settlements. The Mormons could track their stock for quite a ways, but as soon as they got into a sandstone country they gave up the chase, being impossible to track them over the bare sandstone, and no one thought it for a moment possible to get down to the valley below. They never dreamed that here the Indian feasted on broiled steak. Wild oats grow here the same as cultivated does anywhere else but not so heavy. At present is the time when the Indian gathers his yearly supply of seeds and nuts. Those which we left had quite a crop gathered. I have no doubt that our party is the first white party here. The Indians felt surprised how we got in—asked numerous questions—how we found our way in. Wandered about all day trying to find a trail which would lead us out of the canon besides the one we came in by without success. The canon walls are perpendicular from 700 to 1000 ft. in height.

[June] 16. On the search again. At noon Prof. concluded that four of us should climb out and head all the gulches. Prof. and myself went one way while Capt. and Fred went the other. Found a nice level Platteaux on top, but no sign or a trail, but found some very old horse dung, probably three or four years old.

After heading all the side gulches we walked towards the mountains. Here we found ourselves on the divide. We could see the break of the Colorado plainly and only about twenty miles distance. Half the south side of the mountain is drained by a large canon flowing east to c. [Colorado] while the other half flows west and then north into the Dirty Devil River which flows on the north side east into the Colorado. Having studied out our course we were bound to get out of the canon. On our return we climbed down one of the side canons and here we thought we could get up with a little work, but rather hard for the horses, but we were bound to get up. On our return about 4 PM found Capt. and Fred in camp, who reported having seen no sign of a trail, but had found about three miles from where we climbed out, two water pockets, and as our camp was about that distance from the water in the canon, we concluded to try our canon wall. After two hours' work we reached the top. One horse fell backward while going up a steep ledge and fell about ten feet, but picked himself up, shaked himself, and tryed it again—that time succeeded in reaching the top. Camped near the pockets.

June 17, 1872

[June] 17. Got off about 9 AM. Struck out on our trail from yesterday, reached the foot of the mountain [Mount Pennel] about 3 PM. [The] foot hills are heavily studded with Pinone Pine, only found on small streams of water coming down. Climbed up to nearly the top and camped at the head of the spring. Water is very cold. The mountain is ignius rock, Trachyte. No heavy timber grows on it. Fir, aspen, oak (shrub) and pine—both kinds. A very disagreeable day, cold and windy.

June 18, 1872

[June] 18. Remained camped. Prof., Capt. and myself climbed the second mountain [Mount Pennel] from the north end. Fred and Johnson climbed the north mt [Mount Ellen]. After two hours' climbing found ourselves on top. Our Barometer read 13000 feet above the level with the sea, and 3000 above our camp, this being about half way up from the foothills. A grand sight is obtained from this point of the surrounding country. Far to the north is the famous Wasatch Range, while the Sierra La Sal looms up NE from here, and the Sierra Abajo lies due east. Looking south, Mt. Seneca Howland [Navajo Mountain] stands like a sentinel guarding the junction of the San Juan and Colorado, and west is the home of Truthfull James, "Table Mt." The intermediate country between here and all these mountains is cut up by canons and gulches. It seems almost impossible for men to travel over it. While on top a heavy cloud came over, and a snow storm was the consequence. After this had passed we could see another cloud below us—it looked grand. After satisfying ourselves about the locality of the mouth of the Dirty Devil we made our descent. Got to camp in about half an hour. Fred and Johnson returned about sundown.

[June] 19. Started early. Crossed over the divide to the north side between the first and second mountain, and then east. Passed through a forest of cedars. About two o'clock went into camp on the bank of a beautiful little creek [Trachyte Creek] coming down from the second mt. Prof. and Capt. went out to see which was the best way to get into the River, which from this point is plainly visible, but the section of country between here and the mouth of the Dirty Devil is fearfully cut up. Prof. and Capt. returned at supper, having found a trail.

[June] 20. Started off on the trail which led back again to the little creek and followed it down for quite a ways, then taking its course across a Platteaux, which was very sandy. Here we lost all traces of it. We started down the creek again,

June 19, 1872

June 20, 1872

which toward evening wound too much to the south, while our course was east. Went into camp. Prof. and Capt. climbed out but soon returned, having studied out directions, and decided that they would start early tomorrow and find a way in, and if possible to find the gulch in which Prof. had climbed out from the River last fall.

June 21, 1872

[June] 21. Prof. and Capt. started early. Capt. returned about noon—reported having found the gulch, but a Devil's ladder to go down on. Being already, we skinned out, and after three hours travel we reached the head of the gulch [Crescent Wash] which from the top looked dark and gloomy below. After trun[d]ling several pack horses down the cliff, making their hides look like the map of Spain, we finally reached the bottom. Went into camp in a cottonwood grove nearby. Found a pocket of water.

June 22, 1872

[June] 22. Off early. Wound our way down the gulch—reached the River at noon—found it booming and chuck full. After dinner started for the mouth of Dirty River on foot, a distance of three miles. Here we found our boat, the Canonita, all right but the water had washed her sides and floated one oar part way off but lodged between the rocks. We soon cleared her cabins and pulled her out. Soon had her caulked up and went to camp, easier than walking along the cliff. The distance from the water to where we housed her last winter, or fall rather, I estimated at 40 feet, but at present was ten feet below the boat, though it had, as above mentioned, washed the oars, which were buried under her sides in the sand. On reaching camp we hauled out and turned her bottom up, washed her off ready for general repairing and painting .

[June] 23. Caulking boat. Prof., Capt. and Andy started back at 5 PM. I wish them a safe journey back.

[June] 24. Finished caulking and put a coat of paint on her.

[June] 25. Put another coat of paint on her. Fixed up our rations.

[June] 26. Put our rations and other traps in. Started down on swift current—made four miles—stopped to take pictures of an old Moqui Ruin [Figure 6] it stands on a cliff about 15 feet high, 18 feet wide and 20 long.[70] The front, facing north, has fallen in. It must have been hundreds of years ago since it was built. The walls, that is, the stones were honeycombed by age. A good deal of taste was displayed in building. Each room was squarely laid up. As a cement they had used a mortar of mud which was only perceptible at intervals and in thin layers. Camped for the night. Made four views.

[June] 27. Started about 9 AM. Nothing of interest presented itself to photograph. We ran 9 miles and camped Here we photographed more ruins[71] made on one nagitive.

[June] 28. Run down to Pine Alcove Creek—made two views—wind blowing a gale. Found that we had forgotten to bring trypods for dark tent, also focusing cloth was left behind. Rowed to the left bank. Fen. returned. He climbed over

[70] The site at White Canyon (see diary note 22).
[71] The Loper Ruin (see diary note 23).

the cliff, being a cut off. Distance run today 1-¾ miles. Fen. returned about supper time.

June 29, 1872

[June] 29. Made three views, then started down. Went about five miles. Stopped on the right, made two pictures. Had dinner—ran about a mile—stopped —made two pictures of Red Bud Canon. Ran down about a mile and half more and camped at the mouth of creek coming in on the right.

June 30, 1872

[June] 30. Pulled out and ran about two miles and a half, landed on the left. Fred and myself climbed out—found two large pockets about 10 feet wide and 20 deep. Wind blew a perfect gale—could hardly keep on our feet. Laid up for the rest of the day. Could not photograph on account of wind.

July 1, 1872

[July] 1. Went up a side canon—found a large water pocket at the head. Made 9 negatives. Returned to camp and dropped down about two miles. Camped on the left. Two canons coming in—one flows a good sized creek, the other dry, with large pocket at the head. A large island in the shape of a heart is just above camp.

July 2, 1872

[July] 2. Started up the dry canon. On going to the head of the canon found an old decayed Moqui ruin situated under the over-hanging cliff. It was about 70 feet long and 20 wide. I climbed up to it. Found four sticks of oak, but in touching them [they] fell to pieces like Rip Van Winkle's gun. Made two pictures of canon and one of the Island, and two others. Returned at noon. After dinner dropped down about ten miles. Saw a canon coming in on the right. We stopped, Fen and myself took a walk up. I noticed pottery and arrow chips lying around, so we began to look for ruins. We discovered a square hole in the wall of the canon about 12 feet from the ground, but a ledge ran up to it. On examining this closer we found it the work of the Moquis. They had walled up a large cave, leaving a hole in the center two feet square. I entered, found an old corn cob. No doubt they had used this as a store house. Below we could just trace the foundation of three houses. Found the butt of a spear.

[July] 3. Went up and made a picture of the house and two of the canon. Returned, had dinner and pulled out. About a mile down we stopped and made two pictures of a vertical wall, beautifully frescoed. Ran down about two miles more, where Fen. wanted to take a picture but could only get it in the morning. We camped on the left. Fred and myself climbed out. Could overlook the country for miles. Looking northwest, saw a beautiful valley coming in, and also a creek, its edge studded with cotton woods.

July 3, 1872

[July] 4. Glorious Fourth again finds me in the Canons of the Colorado. Fred, as last Fourth, fired a National Salute. Fen and myself made two views, returned at noon—found a layout—a peach pie and coffee cake. It was excellent. Cake light as a feather. I doubt if anybody in the whole United States enjoyed their dinner better than we did, though our cake and pie was baked in a frying pan—but necessity is the mother of invention. Fortunately having two of these gravy makers, Fred made an oven out of the two, turning one on top of the other, in which he put his coals. After dinner, all hands for the top of the canon. Fred sketched in the country, while Fen. and myself made negatives. Succeeded in getting four. Being very hot, we could hardly manipulate. Had to give it up. Returned to camp about 5 PM.

July 4, 1872

| July 5, 1872 | [July] 5. Ran down about three miles and made two pictures. Ran two miles farther to an alcove on the left, made one picture. Had to give it up on account of the high wind. Ran down a few miles more and went into camp. Cleaned glass for the rest of the day. |

July 5, 1872

[July] 5. Ran down about three miles and made two pictures. Ran two miles farther to an alcove on the left, made one picture. Had to give it up on account of the high wind. Ran down a few miles more and went into camp. Cleaned glass for the rest of the day.

July 6, 1872

[July] 6. Made four pictures, then ran down to an alcove on the left. Made four more—ran down to the long cliff bend and camped.

July 7, 1872

[July] 7. Made four pictures, then ran down about four miles and made four more. Ran four miles more, passed the mouth of the big Boulder [Escalante River], but here it flows only from eight to ten inches of water, while on the mountain it just boomed. The sandstone drinks it all. Camped on the left.

July 8, 1872

[July] 8. Ran down half a mile. Made three pictures—packed up and off. Stopped at the junction of the Colorado and San Juan rivers. Made two pictures. About a mile below the junction, camped for dinner. Fred and Johnson tried to climb out but failed. Returned about 4 PM. Ran down to the Music Temple. Could make no pictures on account of light being bad. Monument Canon [Glen Canyon].

July 9, 1872

[July] 9. Rained all day. In camp.

July 10, 1872

[July] 10. Still raining. Cleared up toward evening. Made three views.

July 11, 1872

[July] 11. Finished up and ran down to Mt. Howland [Navajo Mountain]. Stopped and made three pictures. Ran a rapid OK. Had swift water for 15 miles. Camped on the right.

[July] 12. Off early. Ran down to crossing [Crossing of the Fathers]. Dug up a cache of photographic material. Had dinner and pulled out for four hours. Made 25 miles. Camped early on the left. Passed several creeks. One is called "Sentinel" [Wahweap]—a huge column of detached rock stands at its mouth like a sentinel.

[July] 13. Ran down to the mouth of Pah Rio in an hour. Here we expected someone with rations for us, but no one here. Knowing that J. D. Lee had one of his wives [Emma] living here, we started up—found the old gentleman [Lee] in the field plowing. After stating our case, he told us to make his home our home until our men came down, which we accepted. Gave him some flour.

[July] 14. No one yet. Helped Mr. Lee on the road, who started for Jacob's Pools. Willie Johnson went with him.

[July] 15. Hoed onions and beets in the forenoon. Had just returned to the boats when we heard a shout, which afterward proved to be Clem's, who told us that Andy was down the river with the wagon and wanted us to come and help him over a bad place. All went down and helped him up. Read all afternoon.

[July] 16. Fixed up a bridge across the Pah Rio.

[July] 17. Carted over some of our provisions and went to work at the Dean.

[July] 18. Caulked and painted.

July 19, 1872	[July] 19. Painted.
July 20, 1872	[July] 20. Painted.
July 21, 1872	[July] 21. Worked on Lee's dam.
July 22, 1872	[July] 22. In camp reading.
July 23, 1872	[July] 23. Cleaned glass.
July 24, 1872	[July] 24. Mr. Lee invited us to spend the day with him, being the anniversary of their entrance into the Salt Lake Valley. Had a splendid dinner, played cards and sang songs. After supper returned to camp without a change of opinion of Mormonism.
July 25, 1872	[July] 25. Cleaning glass.
July 26, 1872	[July] 26. Made a ground glass for Clem
July 27, 1872	[July] 27. Fennemore feeling very sick.
July 28, 1872	[July] 28. Fixed Clem's tripods and camera box.
July 29, 1872	[July] 29. Hard to work trying to kill time.
July 30, 1872	[July] 30. Doing about the same.

[July] 31. Spent the day with Lee.

[Aug.] 1. Sleeping and reading.

[Aug.] 2. Commenced to make a cultivator for Lee.

[Aug.] 3. Finished it.

[Aug.] 4. Concluded for someone to go to Kanab and find out true state of affairs, and also to report Mr. Fennemore unable to go down the river.[72] Held a counsel to see who should go. Concluded to draw lots. It fell on Andy, but afterward Fred took his place. Clem and myself started up for the horses. Found them seven miles up the canon on the Pah Rio. Returned about 4 PM. I let Fred take my mule—after supper he started, taking with him our mail.

[Aug.] 5. Trying to kill time.

[Aug.] 6. Spent the day with Lee.

[Aug.] 7. Reading and sleeping.

[Aug.] 8. Same as yesterday. Raining.

[72] Fennemore had been sick for some time. In his diary, John D. Lee recorded that Fennemore "now in the employ of Maj. Powell at $100, per Month, became quite feeble through exposure, being of a delicate constituti[on], which rendered him entirely unfit for the laborious duties . . . involving [sic] upon him" (quoted in Brooks, *John Doyle Lee*, p. 312).

August 9, 1872	[Aug.] 9. Saw some geese up the river, pulled over, but when we got there the birds had flown. About 6 PM heard our signal. Hastened down the river. Found Jones, Fred, and Lyman Hamblin with a load of provisions stalled—helped them to make a portage.
August 10, 1872	[Aug.] 10. Received a lot of newspapers. Boys reported the Uintah Utes on the war path.[73] Johnson gone to Salt Lake. Capt. [Francis M.] Bishop driving team for Major. Received a box of candy for the "Boys" from Steward. Three cheers were given.
August 11, 1872	[Aug.] 11. Spent the day reading and packed chemicals.
August 12, 1872	[Aug.] 12. No party yet. Reading and sleeping.
August 13, 1872	[Aug.] 13. Day spent the same as usual. About 5 PM heard our signals. Answered and hastened down. Found the Major, Prof. and wife, Prof. Du Mott [DeMotte][74] and George Adair. Indian Ben for a guide. "Quawgunt" [Kwagunt][75]

[73] Hillers apparently is referring here to the series of incidents and depredations during the spring and summer of 1872 in the San Pete Valley, Utah Valley, and around Nephi and Richfield, Utah (Gottfredson, *History of Indian Depredations*, pp. 294–307). The incidents included Uintah as well as other Ute groups in the area.

[74] Harvey C. DeMotte was a professor of mathematics at Illinois Wesleyan University. Powell had invited DeMotte to accompany him to Kanab during the summer of 1872. DeMotte had agreed to help in determining the exact longitude of Kanab for mapping purposes. Between August 6 and 13, 1872, Powell, Thompson, DeMotte, and others made a traverse across the Kaibab Plateau on their way to the mouth of the Paria River. DeMotte reported his experience in a series of articles in the *Illinois Wesleyan Alumni Journal* in 1873; the articles are reprinted in Elmo Scott Watson, ed., (*The Professor Goes West: Illinois Wesleyan Reports of Major John Wesley Powell's Exploration, 1867–1874* [Bloomington, Ill., 1954], pp. 63–107).

[75] Kwagunt was a Southern Paiute Indian. As a young child, he and his sister had reputedly been the only survivors of an attack (presumably by Yavapai Indians) on his family's band then

In the evening Jones, Fred and myself took Mrs. Thompson and Du Mott boat riding.

[Aug.] 14. Sacked up the rations, put them on the two boats, the Dean and Canonita. After dinner ran the rapids just below the Pahrio. Mrs. Thompson rode over with us. Land just at the foot of it and went into camp.

August 14, 1872

[Aug.] 15. Helped Clem to make a picture then went up to Lee's. Helped to catch horses and mules for the returning party. Signed four vouchers for Major. Had dinner at Lee's Party got started about 2 PM. Mr. Fennemore felt sick. I pity him.[76]

August 15, 1872

[Aug.] 16. Put on extra planking along the keel of both boats. Beaman and [James] Carlton passed here on their way to the Moquis.[77]

August 16, 1872

[Aug.] 17. Major concluded to take only one photographic outfit. Cached Clem's at Lee's, to be brought to Kanab at the first opportunity. On our return to the boats we started down Marble Canon. Walls are very low, but run up very rapidly—grey sandstone at first then the red. We soon reached our old friends the

August 17, 1872

camped on the Kaibab Plateau. The children somehow made their way to another band camped near what is now Kanab. Kwagunt Hollow on the Kaibab Plateau is named for him (see Brigham A. Riggs, "The Life Story of Quag-unt, a Paiute Indian, told to Brigham A. Riggs, a cattleman of Kanab, by the Indian himself," MS on file, Bancroft Library, University of California, Berkeley).

[76] Fennemore returned to Salt Lake City with DeMotte and Francis M. Bishop, arriving there on August 28, 1872, according to the *Deseret News* for August 30, 1872 (quoted in Watson, *Professor Goes West*, p. 105).

[77] See Beaman, "The Cañon of the Colorado."

rapids. Ran over them until we had passed seven. Camped on the left for dinner and pictures. Walls are now from 7 - 800 feet high. After dinner we have a portage then ran four more rapids. Camped on the right at the head of another portage.

August 18, 1872

[Aug.] 18. Made the portage. Made a picture of it from below. Camped for dinner at the head of a rapid on a ledge of rock. No talus. Clem and myself made negatives. Burnt my foot badly. After dinner pulled out the rapid, from that into another which proved to be a portage. After that we had a let down. Camped on the right. Walls 1200 feet—grey and red sandstone with a bedding of limestone.

August 19, 1872

[Aug.] 19. Commenced with running two rapids, then came to a portage. Had dinner at the foot. Made a few negatives. After dinner ran two big rapids and let down three more. Camped on the left. Rained all night.

August 20, 1872

[Aug.] 20. Broke camp 8 AM. Ran a buster of a rapid all OK. Came to a portage, made it, then ran rapid after rapid. Camped for dinner at the head of a big rapid. Made pictures. Large gulch came in on the left. After dinner ran the [illegible] rapid. shipped a good deal of water. Camped early. Made some pictures of a side canon. Have been running the marble up and is now up 100 feet. Walls of canon 2000 feet. The Canonita got full of water, and Prof. not having securely tied the bag in which our negatives were kept got wet and were spoiled—threw them away. Made one portage and ran 12 rapids.

August 21, 1872

[Aug.] 21. Pulled out this morning at 7-½. Ran down a mile and stopped to make a picture of a spring coming out of the side of the canon wall—flowed quite a stream. Also made a picture of a cave. The river, in making a sharp bend, had undermined the wall over 500 feet. Pulled out for a mile or so, then camped for dinner and pictures. Pulled out after dinner. Ran two [of] the biggest rapids we have seen. Canon very narrow, walls 3000 feet, nearly vertical, no foothold for portages. Camped on the left on a sand bank. Ran 10 rapids.

[Aug.] 22. Made two negatives before starting. Ran five rapids. Stopped on the left for a few pictures. Stopped for dinner and pictures on the right bank [Figure 12] Found a field of cáctus apples, very delicious fruit. Have run the marble up. A sort of greenish looking shale is making its appearance. Walls 3300 feet. Vertical on top and talused below. Two gulches coming in from the Buckskin [Kaibab Plateau]. Could see it loom up some 6000 feet heavily studded with timber. Pulled out, ran a rapid, then had a line portage. Kept on running rapids. About 4 PM came to the mouth of the Little Colorado or Flax River, an awful muddy stream and so salt that it cannot be used as a thirst quencher. Camped for time and latitude. This is the end of Marble Canon, 62 miles in length. Ran 63 rapids, made four portages and let down by line five times.

August 22, 1872

[Aug.] 23. In camp. Clem and myself making negatives. Fred climbed out. Prof. and Jones taking observation found a fire place under the rocks, probably built by the Cohonies [Cohoninas].

August 23, 1872

[Aug.] 24. Broke camp and pulled out into the Grand Canon of the Colorado. We began with running a rapid which was followed by four others. When about 6 miles down came to a fissure from which had flowed lava, damming up the river, but the river has cut through it again, only leaving a fall. Went into camp. Major went out to geologise. After dinner Clem and myself went up the river for pictures. Right wall of the canon is 6500 feet high.

August 24, 1872

August 25, 1872

[Aug.] 25. Still in camp. Rowed up the river for pictures. Returned at noon. Made a few more about camp in the afternoon. Major went across the river—found a silver vein in the fissure but how rich he could not tell.

August 26, 1872

[Aug.] 26. In camp. Party climbed out again. Major studying a fault and hunting fossils.

August 27, 1872

[Aug.] 27. Major and Fred climbed for a couple of hours. Prof. filling barometer tube. Broke camp at 10 AM. Ran over the fall without shipping much water. Ran 9 more and went into camp on the left. All hands climbed out except Andy and myself. Jones, Major and Fred rowed across the river and climbed.

August 28, 1872

[Aug.] 28. Pulled out early. Ran five rapids then came to a buster. Let down by line, made it, pulled out, ran another. Landed on the left. Made two fine negatives. Had dinner and pulled out into a rapid, then came to a let down. Ran another small rapid. Came to a young Hell, unloaded the boats. Here it commenced to rain. Being wet from head to foot anyhow, but had to hunt shelter from the beating rain. Got in the river under the side of the boat. Rain over, we hauled the boats over. In doing so I sprained my back. I sunk down, cursed my luck, and crawled to shore. Went into camp—everything wet. Laid down on the wet sand on damp blankets.

August 29, 1872

[Aug.] 29. No sleep last night. Could not raise up. My back seemed broken. Short of rations. Had to do something. Boys helped me up on my legs—after breakfast Fred rubbed me with camphor—felt considerably relieved. About ten AM felt about to walk. Being half way down the rapids and not trusting myself to pull at the oars, I walked down, Andy taking my place,—took the oars at the foot, supported my back. I felt all right. Came to a hell of a looking place. Here the granite comes up, two gulches having emptied their debris, forming the biggest fall we have seen. Stopped at the head, made a picture of it. Had dinner. Major examined it. Could not find a place to let down. Walls on each side vertical, smooth slippery rocks. After dinner pulled into it. In a second found our boat, the Dean, filled in a perfect hell of waves and foam. Got through all right. Bailed out, and watched the Canonita. Only once in a while could see three heads bobbing up, but no boat could be seen. Got through all right. Bailed out and went on into more of not such large calibre. This hell hole has a fall of 60 feet in half a mile. Ran four more, then came to another monster. Concluded to make a portage—foothold being very scarce, had to maneuver very cautiously in order to get at the head of the rapid. Let one boat down to the head of the fall by line, men holding on to the craggs, wedged her in between two rocks at the head of the fall, fastened our line to the stern of the Canonita, and pulled her down all right. Here we unloaded, let down by line over the swift water, passing them around the projecting rocks. Spent all afternoon to make 20 yards. Slept on the bare granite—fine bed for a lame back.

[Aug.] 30. Continued all day hard at work, making a quarter of a mile. About five PM hauled the boats up for repairs. While half through with the job, here came a flood rising the river four feet in an hour and still raining but not with such a rush. Hoisted the boats up and hung them on the wall of the canon, while we went up on the second bench to spend another night on a soft bed of hard granite.

August 30, 1872

[Aug.] 31. River had fallen a little after breakfast. Got everything ready to load her as quick as possible so as not to keep her pounding too long. Everything

August 31, 1872

being ready, we lowered them to the water. In lowering the Dean, she struck a rock which went clean through her, water filling her cabin almost instantly. Nevertheless, could not stop to repair her. Loaded the rations, which are all in rubber sacks, into her cabin, water running out on top. All was ready in a jiffy—we jumped in and soon found ourselves whirling down the tail of the rapid. Landed in a little cove for repairs, and dried out our clothes. After dinner pulled out, which was about 4 PM. Ran ten miles and ten rapids, some of them very dangerous, but shot through them like an arrow. Could not make a portage if we had wanted to. Vertical walls on both sides. Landed at the mouth of Bright Angel River at just 4:45 PM, having made the run in 45 minutes. That is what I call going it, you bet. A lone willow flourishes here—the first since Marble Canon. Camped here on the right. River or creek is on the right.

September 1, 1872

Sept. 1. Pulled the Major across the river. He climbed out but soon returned and started into an old roarer of a rapid three miles long. The Deacon was thrown out but hung to the side and crawled in. Lost three oars but recovered them all below the rapid in an eddy. Ran the granite down and the old red sandstone up. Ran five more rapids. Camped on the right for dinner. Made two pictures. After dinner ran the granite up again. Walls 4000 feet. Ran a big rapid. At 4 PM came to a huge fall, made a portage. Clem and myself made pictures. Climbed up for more. Camped at the foot.

September 2, 1872

[Sept.] 2. Pulled out about 8 AM. Ran into a big rapid. Made it all OK. Ran three small [rapids]. Came to a fall—made a portage—ran a little way—came to another portage. Made it, and had dinner on some rocks on the left. Pulled out after dinner, ran all afternoon, rapid after rapid, until 20 were left behind. Made

138

15-½ miles, 26 rapids and two portages. Camped on the left in a gulch at the head of two rapids, which the Major says we can run, but they are busters.

[Sept.] 3. Pulled out into the rapid. Made the first all right. In the second found hellish big waves, filling us instantly. Got nearly through when she was whirled about by whirlpools. Jones, in trying to hold her head down stream, over-balanced her and capsized, spilling us out. In turning over I made sure of a hold, but got fouled by an oar which knocked me six feet away from the boat, struck a whirlpool, went down head first—down, down, I went. I struggled to turn, finally succeeded. Then commenced a strange sensation. Wind by this time began to tell. I done some of the tallest kicking I ever done in my life. I thought it an age—all at once I felt myself brought up suddenly, and the next instant I had hold of the gunwale of the boat. Major and myself came up together in a boile. Must have been in the same whirlpool with me, as he spoke of being taken down by one, but he fortunately had his life preserver on or else he might have drowned, having only one arm. We soon turned her over and rowed her into an eddy, bailed out and pulled out, landed on the right side at the head of another huge monster. Had dinner, made two pictures. Pulled out into it, made it all right. A mile farther down are some old Moqui ruins on the right, and just below, a rapid two miles long. Major and Prof. examined it. Could make no portage, so had to run it. Pulled into it— while about half through, found ourselves going for the cliff on the right. Pulled hard on the left hand oar. Just missed it by six inches, or else it would have been goodbye, John. Got through all right—ran three more, then came to a young hell. In trying to land at the head on a projecting ledge of granite, Fred jumped out with the line, but missed his mark, fell backward into the river. Major sung, out, Jump boys, jump. He jumped himself to catch the line, but failed. The boat at this time

September 3, 1872

was fearfully near the head of the fall, had swung round, going stern on. In passing a rock close to shore, the Deacon jumped out and just made it. At this time Fred was hanging to the line and in the water. I saw the danger, put on my life preserver, twisted the boat bow on. When I had her turned, I looked around to see where Fred was. I was overjoyed to see his head coming over the bow. In an instant he was at his oars. By this we were at the head of swift water. A huge boulder stuck out in the middle of the river. Got into its eddy. This would break at times, and lucky for us, it broke, sending a large current toward the shore. We took advantage of it, and with might and main, reached the shore. Would have gone over if a friendly rock had not stopped us near the shore. The Canonita was signalled to land, who was let down by line. Camped on the left. Made 3-1/4 miles, 16 rapids, and one let down.

September 4, 1872

[Sept.] 4. Commenced with a let down, then pulled out and ran rapids. Stopped on the left to make pictures, but wind blew so hard had to give it up. Canonita had run down a mile. On arriving at her morrage we found dinner waiting. After dinner, pulled out into rapids. Walls about 4000 feet. Made 14 miles, 23 rapids. Camped on the left, granite down.

September 5, 1872

[Sept.] 5. Pulled out at 8 AM for a little ways. Came to a let down just below. Landed on the left for pictures. Had dinner and crossed over for more pictures, then pulled out, running rapids. Came to the granite again. River immediately narrowed up. At the hea[d] of the granite found the narrowest place on the river, the river not more than 60 feet wide. Landed on the right to examine a rapid, ran it. Half mile farther down heard an angel's whisper in the roar of falling water. Could not get near to see it. Dropped to the head and found it to be a portage. Had to take the boats 200 yards up stream by line over the projecting cliff in order to get over to the left—no foothold on the right. Made the portage and pulled out. Got into the heaviest waves on the tail end of the thing—half filled. Ran more rapids. Ran the granite down and into lava or basalt, properly speaking "trap." Camped on a small bit of sand bank. Made 8-½ miles, three portages, and ran 11 rapids.

[Sept.] 6. Started out with running a huge rapid without shipping a drop of water. Ran three others—landed on the right, for pictures and a portage. Clem and myself made pictures while the rest of the boys let the boats down by line. This is the first rapid that I did not help. At the foot of the rapid a clear, cold stream [Tapeats Creek] comes in on the right—from Buckskin. Went up a ways, found several cascades—made pictures of them. Had dinner. Major went out to geologise. Started down at 3 PM. Ran two old busters. Camped at the foot of the last. Another cold stream comes in on the right—flows a large stream. Made 13–14 miles. One portage and 7 rapids run. Trap all day.

September 6, 1872

[Sept.] 7. Major, Prof, Fred, Clem and myself started up the creek, but had to stop when about a mile up—it began to run in the canon with innumerable falls. Had to climb up 800 feet. Obtained two pictures. A fine view is obtained from this height of the Grand Canon, its walls towering up 6000 feet, and terraced back, each formation forming a terrace. Returned to camp about one PM. Had dinner and pulled out. Ran four rapids in three miles. Saw a creek on the right pouring from a cliff 200 feet high. Stopped and made two pictures. Pulled out and ran nine rapids. Some of the [illegible] ran the granite up and down in three miles. Walls are now all marble again. At 6 PM heard a shout from the mouth of a side canon, which

September 7, 1872

proved to be the Kanab Wash. Found George Adair, Joe Hamblin and [Nathan] Adams with supper for us. Rowed up the creek for half mile to a nice camp. Found one letter from Brother Dick, and lots of newspapers. One letter from Bonemorte. Mrs. Thompson had sent us down some potatoes, cheese, butter and canned fruit. She is the most thoughtful little woman I have ever known. Got in a supply of photographic materials.

September 8, 1872

[Sept.] 8. Cleaned glass and packed up negatives to be sent to Kanab.

September 9, 1872

[Sept.] 9. Quite a surprise this morning at breakfast. Major told us that our voyage of toil and danger was at an end on the river. Everybody felt like praising God. The party would start at noon for Kanab excepting Clem and myself to photograph up the Wash for ten miles, when horses would be sent down to us. Party left at noon. Remained here all day fixing up the traps.

September 10, 1872

[Sept.] 10. Started down the Wash to the river. Made a picture looking up [Figure 13] and one looking [down]. Came back and made a picture of the dismantled boats. Made two more, then went back to camp.

September 11, 1872

[Sept.] 11. Took the traps and worked up the Wash for four miles. Walls of the canon 3000 feet. Lower stratum yellowish looking shale—upper part marble. Returned to camp, leaving the traps behind.

September 12, 1872

[Sept.] 12. Took our blankets and cooking utensils up to where we had left off. Rested, and started up. Made a picture whenever we found good light on it and sometimes would wait for light. Got up about a mile. A stratum of limestone

142

appears on top. Shales gone under, and now marble is on the bottom walls, growing lower. Returned to our blankets and camped.

[Sept.] 13. Took our blankets up to the traps—waited a long time for light— the sun only peeps down into the canon about half an hour, but as the canon changes directions every quarter of a mile, we got good light often. Went up about a mile and a half. Sandstone appears on top. Walls 2000 feet high. Canon in width about 250 feet, which is the general average. Returned to the blankets and camped.

September 13, 1872

[Sept.] 14. Off with the traps. Spent a restless night. This canon is the most gloomiest place I have ever been in—not a bird in it. The only thing of life is the bat and mosquitoes. Made some fine work today. Got up about a mile. Returned to blankets and camped.

September 14, 1872

[Sept.] 15. Started up early. No rest last night. I feel lame and stiff all over. Made fine work today. Lots of cactus apples grow all along the sides of the canon— eat lots of them every day. They are very delicious fruit and I think are very healthy. Indians live on them this season of the year. They also make wine from their juice, which they say makes drunk come. The canon often doubles on itself, leaving only a thin wall in the bend. Photographed one today measuring in height over 2000 feet, all marble, measuring through its base only about 200 feet. Went up a mile, left the traps. Quit early for a picture, which we could not take till morning on account of light. Went back for our blankets, took them up to the shower bath, which I have discovered on a former trip down here. Here we were surprised to find Mr. Adams and Joe Hamblin, who were unpacking for tonight's camp, with ten animals for us, and the caches.

September 15, 1872

September 16, 1872

[Sept.] 16. Adams and Hamblin started down for the caches on the river, while Clem and I went back to make our pictures, then returned to the shower bath. Had dinner and made a picture of it and the canon. Left our traps and started up with our blankets to the head of the water about a mile from the bath, where we camped. Walls now about 1500 (feet). Marble, lime and sandstone.

September 17, 1872

[Sept.] 17. Started back for the traps and made two pictures, when Clem unfortunately broke the slide in the plate holder. Adams and Joe Hamblin returned, having only half executed their orders concerning bringing things, only bringing part of the stuff. Camped.

September 18, 1872

[Sept.] 18. Packed up this morning and started up the canon. Made 20 miles and camped on the left near Moqui Canon. Ran the marble, lime and red sandstone under a kind of red shale and gray sandstone up. Canon half mile wide. No water except a pocket. Walls about 900 feet. The shales[78] are talus. The rest is vertical.

September 19, 1872

[Sept.] 19. Off early, continuing our journey up the Wash. Run the shales and gray sandstone under and limestone up. Made 21 miles. Camped near a pocket under the cliff on the left.

September 20, 1872

[Sept.] 20. On the road early. After riding four miles we climbed out. What a relief to the eye, after being penned up in the canon for three months. Reached Kanab about three PM. Found Mrs. Thompson ready to welcome us. Found an

[78] The words "red sandstone" are written above "shales" in the notebook.

144

actinomy[?] from Dick, also a letter from Dick['s] girls. Fred told us that we were to go to the Moqui towns. Major and Jones gone to upper Kanab.[79] Andy gone to Beaver.

[Sept.] 21. Got ready to clean glass and fix up chemicals.   September 21, 1872

[Sept.] 22. Went to Johnson to see Beaman.   September 22, 1872

[Sept.] 23. Cleaned glass.   September 23, 1872

[Sept.] 24. Clem and Fred taking barometrical observations. Prof. is taking time for longitude.   September 24, 1872

[Sept.] 25. Still to work at glass.   September 25, 1872

[Sept.] 26. Got trypods made.   September 26, 1872

[Sept.] 27. Got a new frame for dark tent.   September 27, 1872

[Sept.] 28. Got a camera box made.   September 28, 1872

[79] Powell and Jones had gone with Chuarumpeak and another Kaibab Paiute man to upper Kanab Creek and Long Valley. From there, Powell, Jones, and Joseph W. Young, a resident of Long Valley, hiked and waded their way down the East Fork of the Virgin River through Parunuweap Canyon to Shunesburg (Jones, "Journal," pp. 160–61). They were apparently the first white men to make the hike.

September 29, 1872	[Sept.] 29. Got through with glass.
September 30, 1872	[Sept.] 30. Andy returned from Beaver. Mrs. Thompson, Fred and Clem went up Kanab Canon to the Lakes, Prof. followed about noon. Party returned late at night.[80]
October 1, 1872	Oct. 1. Fixing breachings for pack saddle.
October 2, 1872	[Oct.] 2. Fixing apparatus. Major and Jones returned from their trip.[81]
October 9, 1872	Oct. 9. Everything was ready for our Moqui trip. Took a wagon to make travel easy as far as the Colorado. Started at 10 AM, Jacob Hamblin as guide and trader. Andy, Clem and myself stopped in front of Johnson's Canon. Took horses up four miles to water. Hamblin, who had remained behind, joined us at Johnson. Purchased several Indian trinkets, hitched up and started for Navaho Wells, where we camped. The well is simply a deep hole in the rock but always plenty of water is found in it.
October 10, 1872	[Oct.] 10. About 4 AM Fred and Charley Riggs rode into camp with a dispatch for Hamblin, stating that Navajos had driven off 75 head of horses and mules from Summit Creek and had taken the southern trail. Jacob and Riggs started for

[80] Thompson ("Diary," p. 100) records these events as occurring on September 29.

[81] There are no diary entries for October 3–8, 1872. W. C. Powell, "Journal," pp. 457–58, indicates the time was spent getting ready for the trip to the Hopi mesas and in taking photographs of the Kaibab Paiute Indians camped near Kanab. Clem also noted that "Jack came to a definite settlement about his wages with him [J. W. Powell]."

the Crossing of the Fathers to head them, which can be easily done. They have to go through a narrow pass, and two men can kill a hundred. Fred after breakfast returned to Kanab, while Andy, Clem and myself drove across the Buckskin to House Rock Spring. Arrived late at night, having come 39 miles. House Rock Valley is quite a long valley, nearly 47 miles long and from two to three miles long [wide?].

[Oct.] 11. Up and off early for the Pools, where we arrived at 2 PM. J. D. Lee has quite an extensive ranch here for his stock. Stopped to water and chat with Mrs. [Rachel] Lee, one of his wives. Found the old gent. had gone to the Pario. The family here lives under a willow shelter, but a stone house is about to be erected. Strove 15 miles farther to a little creek where we camped. Water very alkaline. Burnt up the remains of a gold rocker belonging to some disappointed gold hunter.

October 11, 1872

[Oct.] 12. Started for the mouth of the Pario where we found Jacob who had just arrived. Old John D. met us and welcomed us to his house. Had supper and returned to our old camp of last July.

October 12, 1872

[Oct.] 13. Tryed to raise the Nell which was sunk in the river and buried in the sand. Worked two hours but without success, being wholly under water. Gave it up. Began a skiff and worked hard all day .

October 13, 1872

[Oct.] 14. Still at the skiff.

October 14, 1872

[Oct.] 15. Nearly finished all but pitching.

October 15, 1872

October 16, 1872

[Oct.] 16. Pitched the skiff and put it in the water in the afternoon. Ferryed our traps across about 4 PM. Began to pack up and started [at] sundown. Travel till 9 PM to a creek but failed to find water. Trail very rough at the foot of the cliff.

October 17, 1872

[Oct.] 17. Found water this morning in the creek that is at the head where there is a spring. After breakfast continued our journey at the foot of the cliff. The cliff now bends southward. Traveled on bank 15 miles, then found water in a gulch but so salty it would not boil beans.

October 18, 1872

[Oct.] 18. Started early still going south. Nooned in a little canon where we found water in a little pot hole, but it smelled like manure water. After dinner continued our journey. About sundown turned east. Climbed the cliff, traveled till 9 PM. Found no water for the animals. Jacob took a few canteens, knowing that water was in the neighborhood. He returned with them filled. Had supper and rolled up in blankets.

October 19, 1872

[Oct.] 19. Country around is sandstone. Bare rock curiously eroded forming in good many places. Large pocket(s), found some of them filled. Water our horses. Had breakfast. Clem and myself photographed till noon then packed up and started, taking an eastern trail, leading through a valley then over a low cliff. Night overtook us. Hamblin mistook his direction and another dry camp was the result.

October 20, 1872

[Oct.] 20. Up and off for water which was supposed to be found somewhere. Traveled 15 miles and found water in a large basin.

Had dinner and started again. Made 15 miles more and camped near a butte. Dry camp. Met two Navajoes. *Mise am too wu* bore a recommendation from the agent of the Moquis.

[Oct.] 21. Off early. After traveling five miles met a band of 9 Navajoes—Quinico, an old chief, going to the Mormon settlements to trade. Hamblin is well acquainted with Quinico and they two set up a regular pow-wow. Clem tryed to trade his colt but could not get all he wanted. They offered him four blankets which I think was big. Bid the Navajoes good-bye and traveled on. Found water opposite a large white cliff at S. end in Water pockets. After quenching our thirst we went eight miles farther to the Cutchento Weep (Buffalo Land). Plenty water here — very salt. Hamblin and a party of nine passed through here in '61 on a mission to the Moquis, but were stopped by the Navajoes, then at war with all nations. Their camp was situated on a Table Rock. A mare, the property of George A. Smith, [Jr.,] ran down the trail, Smith following. On turning around a cliff, he was killed by the Navajoes. Three arrows and two bullets pierced his body. Four gray headed Navajoes agreed to pilot them back in safety, who proved to be true. George A. Smith died after surviving two hours. Navajoes scalped him and had a war dance. Party reached settlements in safety.[82]

[Oct.] 22. Made a picture of the place but failed to get a good one—traps out of order. Pushed on till we reached a belt of shrub cedar where we made a dry camp.

[82] This incident is recorded in James A. Little, *Jacob Hamblin among the Indians* (Salt Lake City, 1966), pp. 73–74.

October 23, 1872

[Oct.] 23. Off for Hotoville [Hotevilla] which place we reached at noon. This place is about five miles from Oriby [Oraibi], a Moqui village. A little seep spring keeps alive a dozen little gardens 10×20 [ft.]. Found an old man and two squaws pulling truck. Pushed on to Oriby. Reached it a little after noon. Unpacked and were invited to eat Peakie [piki] and mellon. In the evening Jacob and myself were invited to sup. We accepted. After climbing up a ladder to the second story we were shown a place on a sheep skin. A huge earthen pot set on two stones over a small fire boiling. Directly on a huge bowl was set in front of us. It was filled up from the pot on the fire, which proved to be corn and mutton boiled into a soup. A waiter tray made of willows was next placed on the ground. An armful of cornbread resembling paper cinders rolled up was placed on the tray. Three mellons were brought on from a large stack in the corner. Everything was now ready. Three men and three women seated themselves likewise on a sheep-skin. The old man gave the signal for commencement, by diving with his hand into the bowl of soup—hunted out the biggest piece of mutton. The rest followed suit. Being very hungry myself, and as the old saying is, When you are in Rome you must do as the Romans do, so I sailed in with my digits and pulled forth a dumplin'. Jacob told me not to eat, as it was prepared by the virgin of the house, who had chewed every bit before it was put into the pot. I asked what for. He told me, to arouse the animal passion of the young warior and so hasten her marriage. I allowed the dumplin to roll back, and fetched forth a leg, or rather part of a leg of mutton. I done justice to the meal. I watched the dumplin but none appeared to want it so it was left in the pot. The maiden gave a sigh. While barely through, another Mook [Moqui?] came and invited us to sup. Of course we went, but refrained from eating more soup. Ate a little peakie and mellons. Peakie is corn bread. The meal is made into a thin batter. A large slab of rock is raised about six inches from the ground—

150

a fire built under it. As soon as hot it is poured on—no sooner on than it is done. It is rolled up in wads a foot square. There are three different kinds—red, white and blue (true American) the different kinds of corn. They build their houses on top of the cliff here in Oriby. There are three streets running parallel to each other. Houses are built of stone and clay for mortar—all joining each other, generally two stories high. A ladder is placed on the ground to the top of the first. The second story is set back, allowing a space of eight feet or more to walk on. From this you go into the second story, which has a door. A hole is made in the top of the first and a ladder placed in the hole to get down by. In summer they live in the upper story and winter in the lower. They cultivate the land, raise corn, beans, mellon, peppers and peaches. Raise lots of sheep, asses, a few horses and cattle. Men wear their hair long behind and cut even with the eyes in front. While at work they are naked excepting a breach clout. The women wear their hair long done up in a long roll hanging down on each side. Wear a black blanket dress fastened over one shoulder and a sash—that is, the marryed ones. The marriageable wear their clothes the same. Their hair is done up on the side in the shape of a ram's horn, and are as a general thing pretty—fine features.

[Oct.] 24. Packed up this morning and started for "Haulpie" [Walpi] another Moqui village. After 15 miles travel reached the cliff on which were situated three towns about a quarter of a mile apart. The most southern is Haulpie standing on the end of the maissy 800 feet high. Taba [Hano][83] the most northern and the center

October 24, 1872

[83] The village of Hano on "First Mesa" (the easternmost of the three Hopi mesas) is occupied by Tewa-speaking Indians who migrated to the Hopi mesas in the seventeenth century from the Rio Grande Valley during the Pueblo uprising against the Spanish (Edward P. Dozier, *Hano, a Tewa Indian Community in Arizona* [New York, 1966], pp. 1–10).

is Suny [Sichomovi], where we were shown to a house unoccupied. The occupant had died with the small pocks. We were greeted by two white men, one Mr. Crothers the agent's son, the other a Spanish interpreter, Mr. Wallace. They have quite a stock of Indian goods on hand, but the Indians refuse to take them. Each town wants the whole or none and so they remain stored.

October 25, 1872

[Oct.] 25. Tryed photographing but the traps were out of order. Spent all day fixing them. No better result. Kept on tinkering.

October 26, 1872

[Oct.] 26. Tryed our chemicals but with little better result. A high wind prevailing, gave it up and fixed a new batch.[84]

[84] The entries for 1872 end at this point. Hillers, W. C. Powell, Jacob Hamblin, and the rest of the party remained on First Mesa for several days, trading and making photographs of the Indians and their villages. The party then moved to Mishongnovi on Second Mesa and later returned to Kanab on November 11, 1872 (W. C. Powell, "Journal," pp. 464–71).

On November 30, 1872, Hillers, John Wesley Powell, Walter Clement Powell, Andrew Hattan, Stephen V. Jones, and Joseph Hamblin started for Salt Lake City. Thompson, his wife, and Dellenbaugh remained in Kanab (Thompson, "Diary," pp. 106–107; W. C. Powell, "Journal," pp. 474–77).

In the fall of 1873 Hillers accompanied Powell and G. W. Ingalls from Kanab to St. George and on to Las Vegas to take pictures of the Paiute Indians (see Introduction).

Throughout 1873–74 Thompson, Hillers, and others worked to map the High Plateau area of central Utah (J. W. Powell, *Report of Explorations in 1873 of the Colorado of the West and Its Tributaries* [Washington, 1874]; J. W. Powell, "Survey Under Professor Powell," *Smithsonian Institution Annual Report for the Year 1874* [Washington, 1875], pp. 40–42).

1873

Sept. 11.[85] Left Gunnison at 3 PM. Camped late in Twelve Mile Canon, a mile below saw-mill.

September 11, 1873

Sept. 12. Started for Mooseneah, stopping at saw-mill to grind axes. Camped at noon at the floor of the mountain. After dinner Prof. climbed up, while Joe Hamblin and myself hunted a mule lost in the range.

September 12, 1873

Sept. 13. Started up to the peak with the whole train. Camped right under the nipple. Prof., Joe and myself climbed to the top, taking with us theodolite and barometer. After completing observations started for camp, where we found Robert Duke waiting with a hot dinner. After dinner packed up and started down the mountain on the east side. Camped at the head of Salina Creek.

September 13, 1873

Sept. 14. Pulled out early. Climbed out of the canon. Found several lakes on top. Robert forgot his canteen on top, went back and got lost. Delayed us 45

September 14, 1873

[85] The entries for "Sept. 11" through "Sept. 15" are on a loose sheet, which has no year date. It is apparently 1873, since a manuscript catalog of Hillers's photographs lists several views taken near Gunnison, Utah, in that year ("Catalog of Negatives, River, Land, and Ethnographic, 1871–1876," Bureau of American Ethnology Manuscript Collection, Smithsonian National Anthropology Archives, Washington, D.C.).

minutes. Struck a trail, followed it for some time. Camped at the head of a beautiful valley for dinner. All afternoon in the valley. The trees were dressed in the most gorgeous style. Crossed a large trout stream, but could not stop. Camped near bare bush Butte on a little Creek. Saw three deer. Fired at them but they were too far.

September 15, 1873

Sept. 15. Prof. and myself started for our peak. On arriving at the top found ourselves above the clouds. Waited two hours.

(Subsequent to the first printing, further research has indicated that Thompson's diary ("Diary," p. 118) places Hillers elsewhere on the dates mentioned and suggests that he was still with Powell and Ingalls in southern Nevada. Thompson and a crew including Hillers did work out of Gunnison in 1874, and although Thompson's diary does not cover the dates in question, it is probable that Hillers's entries are from that year and not 1873.)

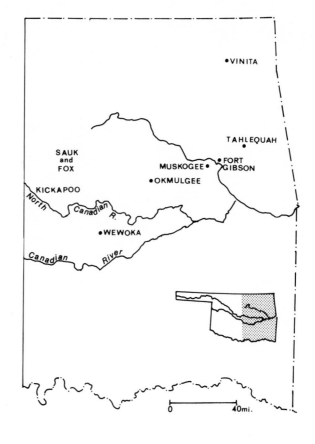

A Map of the Indian Territory, 1875

156

1875

In the Indian Territory

May 1 S[unday] Left Washington, D.C. for Indian Territory.[86] Breakfast at Grafton, dinner at Chilocothe, supper Cincinnati. Scenery very beautiful.

May 1, 1875

[May] 3 M[onday] Arrived at East St. Louis this morning. Crossed Mississippi on suspension bridge. Stopped at the Planters [Hotel]. Received telegram from Major [G.W.] Ingalls.[87]

May 3, 1875

[May] 4. T[uesday] Took a walk through town. Full of business. Mailed a postal order to Mahuken [?] Ingalls arrived in town at noon, his brother and a Mr. Cobble at night.

May 4, 1875

[May] 5 W[ednesday] Took a Pullman for Muskogee. Ingalls took out 40,000 Dollars for the natives. Splendid country for stock-raising. Sceneray very fine up

May 5, 1875

[86] Hillers was assigned by Powell to go to Indian Territory and make a series of photographs to be used in the Smithsonian Institution and Bureau of Indian Affairs exhibits at the Centennial Exposition in Philadelphia in 1876.

[87] Ingalls had previously been an Indian agent in Utah. He, Powell, and Hillers had attended conferences with various Utah Indians in 1873 (see diary note 84).

the Missouri. Changed trains at Vinita on the K. M. and T. Road [Missouri, Kansas and Texas Railroad]. Major [Renfrew M.] Roberts met us at this point.

May 6, 1875

[May] 6 Th[ursday]. Arrived at Muskogee. Small town, a dozen houses, a few stores, rest nigger shanties. I was struck with the intimacy of Negroes and Indians. On inquiry found they intermarry. Fixed up my chemicals.

May 7, 1875

[May] 7. F[riday] Ingalls, Roberts and Doc. [Myron P.] Roberts left for Okmulkee 50 miles from here. My traps being too heavy had to get another wagon, which arrived about 4 PM when I left town.

May 8, 1875

[May] 8. S[aturday]. Arrived at Okmulkee about 7 PM. Found Ingalls, Roberts and Gen. [John Peter Cleaver] Shank[s].[88]

May 9, 1875

[May] 9. S. Fixed my traps and made a picture of Lievers building.

May 10, 1875

M. [May] 10. Photographed Cheyenne Indians [Figures 30, 34, 35].

May 11, 1875

T. [May] 11. Making pictures of Indians.

May 12, 1875

W. [May] 12   "   "   "   "

[88] Okmulkee, or Okmulgee, was designated the capital of the Creek nation in 1867. Beginning in 1870 Okmulkee was the meeting place for the General Council of the Indian Territory which represented all the nations and tribes in the territory (Muriel H. Wright, *A Guide to the Indian Tribes of Oklahoma* [Norman, Okla., 1951], p. 138). General Shanks, a Special Commissioner for the Bureau of Indian Affairs, addressed the Council on May 5, 1875 (Journal of the Sixth Annual General Council of the Indian Territory, p. 12 [copy in the Indian Archives Division, Oklahoma Historical Society, Oklahoma City]).

Th. [May] 13. Pho[tographed] the council and Tocopes.[89]

F. [May] 14. Phot[ographed] the Pawnees [Figure 32].

Sat. [May] 15. Left for Weewaukee [Wewoka].[90] Camped at Ike Smith's, 15 miles from Okmulkee. Mr. Brown, a half breed Seminole accompanied me.[91]

S. [May] 16. Started early. Crossed many prairies. Fine stock country. Arrived at Wewaukee about 7 PM. Stopped six miles east of Wewaukee and made two pictures of cascades. Very beautiful. Found the river up and booming. Crossed on a large sycamore which had been felled across the stream. Stopped at Mrs. Lilly, a widow who keeps a hash house, and is assisted by her daughter and sister.

[May] 17 Made pictures of council house and church.

[89] It is not clear what "Tocopes" means.

[90] Wewoka was established as the capital of the Seminole nation in 1868 (Wright, *Guide to the Indian Tribes*, p. 235).

[91] This is probably John F. Brown, known as Governor Brown. He was the son of a Scottish physician, Dr. John Frippo Brown, and a Seminole woman. Dr. Brown had attended the Seminoles during their move from Florida to Indian Territory. The son, John F. Brown, was a partner in the highly successful Wewoka Trading Company (Wright, *Guide to the Indian Tribes*, pp. 234–35). Hillers's references to Brown's "English nobleman" father and his being a "trader" strengthen the inference that this is "Governor" Brown. One source, however, indicates that Dr. Brown was a "South Carolinian of Scottish descent" (Mrs. William S. Key, "Tribute to Alice Brown Davis." A copy of this address is in the Indian Archives Division, Oklahoma Historical Society, Oklahoma City.).

May 18, 1875	[May] 18. T. Ingalls and Major Roberts left this morning, also General Shanks, with the Cheyenne Indians in charge of Phil McCusker, interpreter for the Wichita Agency.[92] Made a picture of John Chupco,[93] Factors church and a family group of Mr. Brown's.
May 19, 1875	W. [May] 19. Started for Col[onel John] Jumper's, Chief of Seminoles,[94] 15 miles. Made a picture of his church. Started back about a mile to the trader, a Mr. Brown, half breed, whose father had been an English nobleman. I was show[n] all his valuables which are still preserved. He was or had been Prof. of Edenbourgh [Edinburgh] College.
May 20, 1875	T. [May] 20. Started back for Weewaukee. Made a picture of Brown's house then came 15 miles and camped at a Mr. [E. A.] Eggleston, a rancher and trader. A storm came up, the worst I have ever seen. It lasted for ten hours.
May 21, 1875	F. [May] 21. River high. Could not cross Little Wewaukee. Took a ride downstream. Came to a house, found a pair of twins spread out on a blanket, naked

[92] Philip McCusker was a white married into the Comanche tribe who was regarded as its official interpreter (A. A. Taylor, "Medicine Lodge Peace Council," *Chronicles of Oklahoma* 2 [1924]: 106–107).

[93] John Chupco (d. 1881) was the leader of the "loyal" faction of the Seminole Indians, i.e., the group that sided with the Union during the Civil War (Wright, *Guide to the Indian Tribes*, pp. 234–35).

[94] John Jumper (d. 1896) was leader of the faction sympathetic to the Confederacy in the Civil War. He organized the Seminole Battalion of the Confederate Army, achieving the rank of colonel (Wright, *Guide to the Indian Tribes*, pp. 234–35; Carolyn T. Foreman, "John Jumper," *Chronicles of Oklahoma* 29 [1951–52]: 146–47).

in the shade. They looked well, and judging from the mother's fulness they were supplied with plenty of clabber. What would I not have given for a photograph of all, but I had not the collodion to do it with.

S. [May] 22. Started off this morning, the creek being fordable. We went five miles then to be stopped by the high water in the North Fork of the Canadian. Camped with an Indian who was a poligamist, having two wives. He treated us first class. Dr. Robert, correspondent for the Inter Ocean, and a nigger, a lazy brute.

May 22, 1875

S. [May] 23. Started across the North Fork of the Canadian River. A lot of Indians crossed the same time going to camp meeting. Arrived at Okmulkee at 6 PM.

May 23, 1875

M. [May] 24. Started for Muskogee. Joe Cornells here took his own team. Had a lady passenger weighing about 250, a half breed at that. Camp 25 miles out.

May 24, 1875

T. [May] 25. Arrived at Muskogee about 11 AM.

May 25, 1875

W. [May] 26. Started for Tahlequahm [Tahlequah], the capitol of the Cherokee Nation. Arrived at 7 PM. Stopped at the Rev. J[ohn] B[uttrick] Jones.[95]

May 26, 1875

T. [May] 27. Made pictures of house and family and three young half breeds.

May 27, 1875

[95] Jones was a Baptist missionary and had earlier been an Indian Agent to the Cherokee in 1872–73 (*Report of the Commissioner of Indian Affairs for 1873*, p. 294).

May 28, 1875	F. [May] 28. Went to Hildebrand Falls, seven miles from T[ahlequah]. Mr. Stevens and Williams accompanied Gus and myself. Made the pictures and returned by 7 PM.
May 29, 1875	S. [May] 29. Started for the Ladies Seminary[96] with same party. Made pictures of all the pupils. Quite a lot of pretty girls. Started for the Orphand Asylum, but too late to make a picture, as the sun was in the camera. Returned to town and made picture of council house.[97] Returned to Jones'.
May 30, 1875	Sunday [May] 30. Fasted.
May 31, 1875	M. [May] 31. Returned to Ft. Gibson. Made picture of Asylum and one cascade.
June 1, 1875	T. [June] 1. Made picture of Chief Ross' house and one of himself.[98] Returned to Muscogee.
June 2, 1875	W. [June] 2. Made arrangement to go with Major [John H.] Pickering to Sac and Fox Agency.[99]

[96] A photograph of the seminary, not by Hillers, is contained in Wright (*Guide to the Indian Tribes*, p. 71). The seminary for women, a second seminary for men, and the orphan asylum at Fort Gibson were built and operated by the Cherokee nation (S. W. Marston, "Report of the Union Agency, I.T. to the Commissioner of Indian Affairs, for 1875–76," *House of Representatives, 44th Congress, 2nd Session, 1876–77, Executive Documents* I, part 5, p. 465).

[97] ". . . now the [Cherokee] legislature assembles in a spacious brick council house . . . which cost in erecting the sum of $22,000" (Marston, "Report of the Union Agency").

[98] William Potter Ross (b. 1820) was elected principal chief of the Cherokee in 1872. He was the nephew of Chief John Ross who led the Cherokee from 1828 to 1866 (Grace S. Woodward, *The Cherokees*, The Civilization of the American Indian Series no. 65 [Norman, Okla., 1963], p. 318; Wright, *Guide to the Indian Tribes*, pp. 62–72).

[99] Pickering was the Indian Agent at the Sac and Fox agency.

Th. [June] 3. Started this morning—two wagons. Pickering, wife and adopted child, and a Prof. Pickett, clerk for Pickering. Mr. Whisler [John Whistler], a rancher,[100] took my traps. Gus and myself reached Okmulkee about 7 PM. Had a good time with Mr. Austin,[101] the ever-attentive bookkeeper for Mr. Lievers. After lubricating from a flask of brandy we had supper, on which occasion Billy was quite flowery. Topic, calling of man.

F. [June] 4. Started this morning. A storm came on while a few miles out from Okmulkee. Stopped for dinner about 15 miles out. About four PM came to a creek which was up and booming, which prevented the wagons from crossing. The Pickerings crossed on a foot log and started for a ranch, mile and a half farther on. Whisler, Gus and myself remained. Storm came on about midnight. The creek, which had gone down by 11 PM now raised again.

S. [June] 5. Had to wait till 2 PM, when we crossed. Water touched the wagon box and a little went in and wet my glass, damn it. An hour's drive brought us to deep fork. Here we went across on a foot log in search of Pickering who had come down to the crossing. We unloaded our wagon and brought it to the opposite side on the foot log, then attached a line to the wagon tongue, which an Indian brought across by swimming. Camped at night at a Mr. Butler, a Negro-Indian.

S. [June] 6. Started this morning early. Raining nearly all day. Arrived at the Agency about 3 PM.

---

[100] Whistler was also a trader at the Sac and Fox agency.

[101] Possibly this was W. L. Austin, who is listed as a trader from late 1875 to 1877 in the Sac and Fox agency records (Rella Looney, personal communication).

June 3, 1875

June 4, 1875

June 5, 1875

June 6, 1875

June 7, 1875

June 8, 1875

June 9, 1875

June 10, 1875

M. [June] 7. Fixed my traps and started for the Mission House to make a picture but traps would not work, so returned and fixed up for tomorrow.

[June] 8. T. Wind blowing like hell. Made two large and three stereos of Agency and Mission House.

[June] 9. W. The Indians held a council. Photographed 3 women and two men. Stereo. out of fix.

[June] 10. Th. Started this morning for Kickapoo towns with Gus and the interpreter. When seven miles from towns broke wagon tongue, which was spliced at the ration house. Quite a number here drawing rations.

1875

A Letter

Dear Brother:

While I have time I will spin you my log. On my arrival at Muskogee I was greeted by a sight little expected—a big burly Indian kissing a Negroe wench. On inquiring found that the Indians and niggers intermarry. This is just the contrary among the wild tribes. They hate a nigger and think him a being of the devil, an Unupits, but only the Creek Indians believe in the amalgamation of red and black. In the Cherokee where they are nearly all half breeds it is forbid[den] by law and a prison offence. I spent the day at Muskogee, and on the following day set out with an old darky for Okmulgee, 80 miles inland. Major Ingalls had preceded me in his carriage. About 7 PM my King of Sheba arrived for my utencils. His team consisted of two mules whose sides were fastened to a single straight gut without a stomach, and an old fashioned home-made wagon. At the sight of these animals led me to ask how soon he expected to make Okmulgee. He assured me that he would land me at that place tomorrow's sundown. Off we started. I found the old man very inquisitive, but like the majority of Negros, very ignorant, and believed in the Baptist doctrines, but I think the old man was terrible puzzled concerning his future state. After we arrived at Okmulgee on the time promised by him, I bid him goodbye. Here I found Major Ingalls arranging matters. The

Grand Council was in session and all Nations were represented living in this territory. I grouped the whole and made a No. 1 negative. Here I found six Cheyennes who had just left the war path, all strappen big fellows. I took them among the rocks and set them up as food for my camera [Figures 30, 34, 35]. I stripped them to the buff, not a stitch on them except a breach clout [Figure 35] and succeeded in making pictures of them all. I spent just a week here packing up odds and ends. From here I went to Weewaukee, Seminole country. You see, Dick, everything ends with a *kee*, but damn it, no whiskey. The half breeds and bloods find a substitute in "Jamaica ginger", "Pain Killer", Mustang liniment, and "Rahways' Ready Relief," and a little of it converts them into the once savage "plume taker." I engaged a man to take me over, but owing to sickness he was prevented from going, so he let me have his mules, two good animals, and a spring wagon. Council had adjourned and Mr. Brown, a half breed member of the council for the Seminoles, returned home with me. On our way found two little cascades, which I photographed. After a day and a half's journey we arrived at the village, consisting of half dozen houses or more, blacksmith shop, saw and grist mill, the latter owned and run by a white trader named Long. On arriving found Ingalls, General Shanks, and Major Roberts had arrived. Stopped at the house of Mrs. Lilly, a widow. She runs a hash house, assisted by her sister and a daughter, a beautiful Lilly of sweet 20 or thereabouts. I got along nicely with the young miss, wheather it was to make herself amiable, or simply to work herself into my good graces for a picture of herself, I was in doubt. However, I pleased her. She made some attempt at singing and playing an accordion, but neither was first class. After staying a few days here I went to the home of that renown[ed] chieftain, old John Jumper, who ten years ago counted his scalps by the hundred, and who then wore a buckskin shirt trimmed with the hair of his victims, but now a peaceable farmer and expounder of the Gospel. Liken to Paul, he had seen a vision, buried the tomahawk and drove the scalping knife into a large sycamore. Arriving at his home found him tending to his children. He formerly was a poligamist and had three wives, but when he joined the church he was told to let two of them go and retain only one. He kept the children. What became of the women I could not learn. Close by his house he has erected a large building 10×60 [ft.] which is used as a school house and church. Quite a bell is suspended in the cupelo on top. A half breed Choctaw lady is the teacher of the young ideas. Now, Dick, I must describe to you this really beautiful country. Ride with me in mind: The road to this Col. Jumper's place from Wee-waukee winds through low foothills through which a hundred little streamlets wind their way to the Canadian River. Their waters are clear and sparkling, and as they tumble over some moss-covered ledge it is dashed into foam. The banks are fringed with foliage equal to the tropics. Here the stately oak and pecan tree, with their rich garb of bright green, luxuriate, and their branches form a lattice work for the wild grape, honeysuckle and ivey. The prairie is covered with wild flowers of every hue and the whole air is perfumed with [their] sweet balm, especially in the morning, and a million birds greet the sweet dawn of day. And at midnight, if camped in some beautiful grove by some trinkling spring, your dreams are disturbed by the sweetest singing of the nightingale. What a dreamland this is. It is a garden spot of America. But soon I must exchange this for the sage brush, sandy plains, and alkali water. I, on my return to Muskogee, took a trip into the Cherokee Nation. I went to Tahlequah, the capitol of the Nation. Mostly half breeds reside here and some are so white that you cannot tell them from the bona fide white. The young misses are exceedingly beautiful. They retain the fascinating black eyes and glossy black hair of the Indian. I went to the Ladies' Seminary to photograph this piece of Indian enterprise. Being well backed by letters of introduction from prominent men of Tahlequah, I started for the Institute, which is situated about three miles south of the town. On my arrival, was kindly cared for by the principal teacher, a young miss on the yellow-leaf side of thirty, who introduced Prof. Hillers, Govnt. Photo. to her young and beautiful half, quarter, and octavo blood Indians. I soon made known my wants, stating that I desired a picture of their buildings and their beautiful selves, for to be put on exhibition at the Centennial. Soon I saw them arrayed in their finest, and really, their dresses would not be sneezed at in any metropolis. A very beautiful quarter-blood came to me after my grouping had been completed, and with the sweet bewitching smiles, asked me to place her in position. Of course, I would, and show her beautiful form such as but few women possess, to an advantage. At the sight of her large lustrous black eyes and long jet black hair and her regular cut features, and a smile so captivating, threw me almost into cupid's arms, and for a moment I forgot my bachelor's resolution. I placed her near the center. A few small stones were lying loose on the ground. I told her to be seated, three-quarters front. I placed her delicate little No. 2 slipper foot on a small stone, her head resting lightly on three fingers, her eyes turned up and a sweet smile spread all over her face. Her dress of white and a black lace thrown over her shoulders, she set like a queen. I made the picture which was truly a good one, and how it happened I could not tell, for my nerves were all unstrung, my hand trembling, I could scarcely handle the plate. After they had all seen it and admired it, they returned to their rooms. What had become of my queen? I could see her nowheres. I felt sick. I packed up, took my picture into the kitchen to varnish it by the fire. Another crowd came rushing in who wanted one more look. Of course I gratified their desire. And I wished my Ideal would come in and have a look — not that I cared to have her look, but I wanted to feast my own eyes. I was doomed to disappointment. I bid them goodbye, mounted my horse and sadly rode away.

The landscape so beautiful about the Seminary, had no charm for me — my thoughts were of the beautiful Indian maid. I asked myself if this could be anything like love. Well, I concluded, if this painful desire was called love, I wanted to be rid of it, and that very soon. I changed my thoughts to the days when I was tempest-tossed on the Colorado River, shooting over falls and rapids, climbing pinnacles and towers, and rushing rock-ribbed gulches — a life full of adventure. While in this revery I heard the sound of a horse's hoofs and the breaking of limbs. Involuntarily my hand went to my pistol holster, pulled it out and looked to the priming and before I could replace it a sweet voice entreated me not to shoot her. If, no doubt, could I have seen my face I would have been surprised at the many colors. Mounted on a thoroughbred sat my Vision. The color had deepened on her face from the exercise and looked more beautiful than ever. She begged pardon for her intrusion, but she thought of going home and as my road was nearly in the same direction she thought of placing herself under my protection (thanks for your good opinion). We rode about two miles when my road led in an opposite direction. She told of a cascade which was situated near her home, which induced me to change my mind. As it was early I could have plenty of time to return before night. So I stated to her that I would go with her and see the falls. She seemed pleased. We talked of everything, about their beautiful country, people and music. We arrived at her father's house early afternoon. I unpacked, while she oppened the bars to allow my animals to go into pasture. I asked her motive. She responded that my animals would be safe in there for the night. I said No, I must go back to the city. All right, we will see about that. She called a helper to have my blankets and traps stored away until wanted. Just imagine your poor old brother for a moment, dressed in a gray shirt with large collar, black silk sailor's scarf tied with studied negligence, broad sombrero, a pair of 5 dollar cowhide boots

169

which were adorned at the heels by a pair of Mexican silver spurs, pants inside, full grown beard and long hair. In this style my maid led me to the house, which set back in a yard full of flowers. The first we met was the Colonel, her father whom she introduced me to as a Jerome. The old gent was pleased to have someone with whom he could talk. The old lady was a pleasant old lady, rather fat and a half breed. The old man talked politics, and a rebel sympathiser, an argument which I studiously avoided. A lunch had been set out for my Alvoretta and myself. After a thorough absolution from dirt I sat down and enjoyed a first class lunch. She had read Scribner's Magazines and had seen my pictures in them,[102] which of course she praised to the skys, and no doubt she meant. I proposed a walk to the falls, which of course the Colonel would not listen to, not today — tomorrow. But, my dear sir, I must go back to the city. But don't mention city — you're my guest for the night and the other nights while you stay in the Nation. Just make yourself at home. Here, Al, play some music for the Professor, to get him out of the notion of going to town. I was very easily persuaded. The house was elegantly furnished, the parlor was superb, an 800 dollar piano graced one side of it. I was thunderstruck when my vision began "Ever of thee I (am) fondly dreaming." I think my hands trembled while I turned the leaves for her. Her sweet music and song caryed me back to days gone by when I was courting a young widow down south in Georgia, but such is life (!). She begged me to sing, which I did and then sang a duet, "What are the wild waves saying," she was so pleased that we sang it twice over. The old man was a great smoker and seemed to enjoy to make smoke rings while so enjoying

[102] Hillers is referring here to the engravings based on his photographs used to illustrate three articles which John Wesley Powell did for *Scribner's Monthly* in 1875–76. The first two articles may have been available by May 1875. See John Wesley Powell, "The Cañons of the Colorado," *Scribner's Monthly* 9 (1874–75): 293–310, 394–409, 523–37, and, "An Overland Trip to the Grand Cañon," *Scribner's Monthly*, 10 (1875): 659–78.

ourselves on the varanda, the Colonel's chosen place for the enjoyment of his Havanna. I heard the quavering chords of a guitar in the garden below playing, "Come love, come, ere the night torches pale," etc. As the last sweet note had died away the Colonel proposed to retire, which I also was pleased to do. As I slept long this morning, something very unusual with me, found my host waiting breakfast for me, a beautiful boquet of flowers lay by the side of my plate. I thanked the unknown donor. Breakfast over, we started for the falls. The Colonel feeling somewhat indisposed, excused himself, so I had my queen for a companion and guide. The road to the cascade let through an oak grove.[103]

[103] The diary ends abruptly at this point and we have no further indication of the outcome of this *tête-à-tête*.

Epilogue

The photographs Hillers took in Indian Territory in 1875 were used in the Smithsonian Institution exhibit at the 1876 Philadelphia Centennial Exhibition.

In later years Powell sent Hillers on several expeditions to the West for the Bureau of Ethnology and the Geological Survey. One of the more important of these was an expedition to the pueblos of New Mexico and Arizona and to Canyon DeChelley, Arizona, in 1879 (Figures 36–43). Hillers was accompanied by James Stevenson and Frank Hamilton Cushing.[1]

Hillers was chief photographer of the Geological Survey until 1900. He continued working for the Survey on a part time basis until 1919 when ill health and advancing age forced him to retire.

Hillers's career as a professional photographer spanned a period of nearly fifty years. It began through a chance meeting of the young teamster Hillers and the young scientist Powell in 1871 in Salt Lake City. Through his research and administration of research Powell came to have an enormous influence on the courses of American anthropology, geology, and conservation practices.[2] Hillers, through

[1] Raymond S. Brandes, *Frank Hamilton Cushing: Pioneer Americanist*, Ph.D. diss., University of Arizona, Tucson (University Microfilms 65–9951).

[2] William C. Darrah, *Powell of the Colorado* (Princeton, 1951); Wallace Stegner, *Beyond the Hundredth Meridian: John Wesley Powell and the Second Opening of the West* (Boston, 1954).

his photography, played a vital role in documenting the researches of Powell and his associates. On another level Hillers's photographs remain a monument to his skill and ability. He was truly one of the great photographers of the nineteenth century American West.

Jack Hillers's Photographs
of the Powell Expeditions,
1871–1875

1 Second Powell River Expedition, May 22, 1871, at Green River Station, Wyoming Territory. Left boat (l–r), E. O. Beaman, Andrew Hattan, Walter Clement Powell; center boat (l–r), Stephen Vandiver Jones, John K. Hillers, John Wesley Powell, Frederick S. Dellenbaugh; right boat (l–r), Almon Harris Thompson, John F. Steward, Francis Marion Bishop, Frank Richardson. Photographer unknown.

All photographs are those of Jack Hillers unless otherwise noted.

179

2 Ashley's Falls, Green River.
Note boat at left. E. O. Beaman photograph.

180

3 Harrel's Party, Brown's Park, Green River, 1871.
E. O. Beaman photograph.

4 Powell expedition members
at Dodd's cabin, confluence of Uintah
and Green Rivers, 1871. 1. F. S. Dellenbaugh,
2. J. K. Hillers, 3. A. H. Thompson,
4. E. O. Beaman (?), 5. W. C. Powell,
6. J. F. Steward, 7. A. J. Hattan,
8. F. M. Bishop. Photographer unknown.

5 Boats in Desolation Canyon, 1871.
Probably J. F. Steward seated in
"Nellie Powell." Note boat being "lined"
through rapids at left. E. O. Beaman
photograph.

7 Three Patriarchs,
Zion Canyon, Utah, 1872.

6 "Moqui" ruin on ledge, confluence of
White Canyon and Glen Canyon, 1872.
J. Fennemore and J. Hillers photograph.

8 Zion Canyon, Utah, 1872.

9 Kolob Plateau, Utah, 1872.

188

10 Inner Gorge of Grand Canyon, Arizona, probably 1872.

189

11 Marble Canyon,
Shinumo Altar in foreground, probably 1872.

12 Marble Canyon, Arizona, 1872.

13 Marble Pinnacle, Kanab Canyon,
Arizona, 1872.

14 "Summer Home Under a Cedar Tree."
Chuarumpeak, leader of the
Kaibab Paiute band is second from the left.
Near Kanab, Utah, 1873.

15 "Mother and Child." Kaibab Paiute near Kanab, Utah, 1872(?). Note the museum accession number and the word "Colorado" on the bodice of the woman's dress.

16 "The Tavokoki or Circle Dance."
Kaibab Paiute near Kanab, Utah, 1872(?).

17 Three Paiute women in native dress.
Near Las Vegas, Nevada, 1873.

18 "Wu-na-vai Gathering Seeds."
Moapa Valley, Nevada, 1873.

19 "Chuarumpeak Shooting a Rabbit."
Kaibab Paiute men near Kanab, Utah, 1873.

20 "Ta-noats." Las Vegas Paiute.
Las Vegas, Nevada, 1873.

21 "E-nu-ints-i-gaip, an old man."
Las Vegas, Nevada, 1873.

22 "Won-si-vu at Rest." Kaibab Paiute
on Kaibab Plateau, Arizona, 1873.

24 J. K. Hillers, probably near Kanab, Utah, 1872 or 1873.
Photographer unknown, possibly J. Fennemore.

25 High Falls, Bullion Canyon, Utah, 1874.

23 J. K. Hillers sewing. Red Canyon Park, Green River, 1874. Photographer unknown.

26 Pilling's Cascade, Bullion Canyon,
Utah, 1874.

27 "The Mirror Case." John Wesley Powell
and Uintah Indian Woman,
Uintah Valley, Utah, 1874(?).

28 "Sai-ar and His Family." Uintah Valley, Utah, 1874.

29 "A Domestic Camp Scene." Uintah Valley, Utah, 1874.

30 "Plenty Horses, a Cheyenne."
Near Okmulgee, Indian Territory, 1875.

31 "Pah-ri-ats in Native Winter Dress."
Uintah Valley, Utah, 1874.

32 "A-sa-wa-ka-red-i-hewl,
or Big Spotted Horse," a Pawnee Indian.
Near Okmulgee, Indian Territory, 1875.

33 "Big Mouth, or Binanset," an Arapaho.
Near Okmulgee, Indian Territory.

34 "Little Bear, a Cheyenne."
Near Okmulgee, Indian Territory, 1875.

35 "Feathered Wolf, a Cheyenne."
Near Okmulgee, Indian Territory, 1875.

37 Ceremonial Dance at Zuni Pueblo,
New Mexico, 1879.

36 Zuni Pueblo, New Mexico, 1879.

Zuñi maiden.

39 Canyon DeChelly, northern Arizona, 1879.

38 "Tsa-wel-la-tsi-ta, a Girl of Zuni, New Mexico." 1879.

41 "Terraced Houses,
Oraibi Pueblo, Arizona," 1872 or 1879.

40 "A Court View of Oraibi Pueblo,
Arizona," 1872 or 1879.

42 "Hopi Sub-chief." Hopi Pueblos,
Arizona, 1879.

212

43 "Hopi Man Using Spindle Whorl."
Walpi, Hopi Pueblos, Arizona, 1879.

213

44 Portrait of John K. Hillers, 1880s.
Photographer unknown.

Index of Personal Names

Index of Place Names

225